Evander Luther (Ed.)

Samsung Galaxy Note

Evander Luther (Ed.)

Samsung Galaxy Note

Android (operating system), Smartphone, Tablet computer

Acu Publishing

Publisher:
Acu Publishing is a trademark of
International Book Market Service Ltd., 17 Rue Meldrum, Beau Bassin, 1713-01 Mauritius
Email: info@bookmarketservice.com
Website: www.bookmarketservice.com

Published in 2012

Printed in: U.S.A., U.K., Germany. This book was not produced in Mauritius.

ISBN: 978-620-0-46830-7

Contents

Articles

References

Samsung_Galaxy_Note

Manufacturer	Samsung Electronics
Compatible networks	(GSM/GPRS/EDGE): 850, 900, 1800, and 1900 MHz UMTS: 850, 900, 1900, and 2100 MHz HSPA+: 21 Mbit/s; HSUPA: 5.76 Mbit/s; LTE
Related	Samsung Infuse 4G Samsung Galaxy S II
Type	Touchscreen smartphone
Dimensions	146.85 mm (5.781 in) H 82.95 mm (3.266 in) W 9.65 mm (0.380 in) D
Weight	178 g (6.3 oz)
Operating system	Android 2.3.6 (Gingerbread) with TouchWiz UI 4.0
CPU	1.4 GHz dual-core ARM Cortex-A9 SoC processor; Samsung Exynos 4210 (GT-N7000) / 1.5Ghz Qualcomm Snapdragon 8255T (GT-N7003)
GPU	ARM Mali-400 MP (GT-N7000)/Adreno 205(GT-N7003)
Memory	1 GB RAM
Storage	16/32 GB flash memory
Removable storage	microSD (up to 32 GB)
Battery	Li-ion 2500 mAh
Data inputs	Multi-touch touch screen, headset controls, proximity and ambient light sensors, 3-axis gyroscope, magnetometer, accelerometer, barometer, aGPS, GLONASS and stereo FM-radio, S Pen (Stylus)/Pen UX
Display	1280x800 px, 13.46 cm (5.3 in) at 285 ppi WXGA HD Super AMOLED
Rear camera	8 Mpx 3264x2448 with auto focus, 1080p 30fps Full HD video recording, and stills. Single LED flash.
Front camera	2 Mpx for video chatting, video recording (VGA), and stills
Connectivity	3.5 mm TRRS; Wi-Fi (802.11a/b/g/n); Wi-Fi Direct; Bluetooth 3.0; Micro USB 2.0; Optional Near field communication (NFC); USB Host (OTG) 2.0
Other	Exchange ActiveSync, integrated messaging *Social Hub*, *Readers Hub*, *Music Hub*, *Game Hub* and *Samsung ChatOn*

The **Samsung Galaxy Note** is an Android smartphone and tablet computer that was introduced in October 2011. It has attracted attention because of its 5.3-inch screen size – between that of conventional smartphones and tablets – and because of its included stylus.

History

The Galaxy Note was announced by Samsung during IFA 2011 in Berlin. It was released to the public starting with Germany in late October 2011, with other countries following shortly after that.[1] By the end of November, it was available in most major markets, including East Asia, Europe and India.

In December 2011, Samsung announced that 1 million Galaxy Notes were shipped in less than two months, and that Galaxy Notes will be available in the US in 2012.[2] In early January, advance units, customised for the AT&T carrier, were supplied to the media, in preparation for its public release in the United States market.[3]

Size and construction

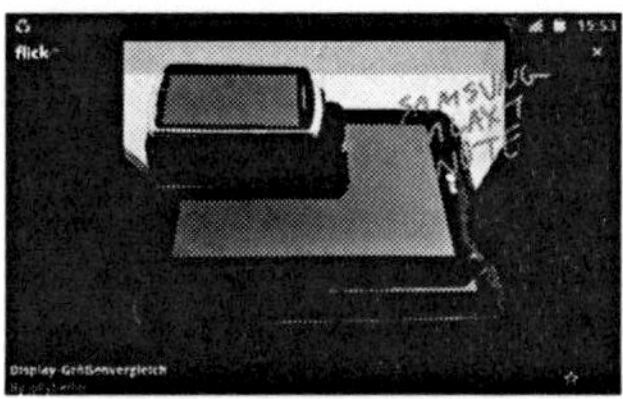

A screenshot annotated on a Galaxy Note, showing a picture comparing the Galaxy Note's size with other smartphones and tablets (Order from top to bottom: Sony Ericsson Xperia Play, Samsung Galaxy S II, HTC Titan, the Samsung Galaxy Note itself, T-Mobile G-Slate, and Samsung Galaxy Tab 10.1).

The Galaxy Note, with a 135 mm (5.3 in) screen, occupies a niche in Samsung's Galaxy range, bridging the gap between their smartphones such as the Galaxy S II, and the Galaxy Tab tablet. Reviews regarding its size have been mixed. Some have said that it's too big to be a true mobile phone, citing the difficulty of using it single-handed [4], storing it in a small pocket, and the perceived ungainly appearance against the face [5] ; while others have hailed it as a pioneer in a new market segment (despite the lack of success of the similarly-sized but stylus-lacking Dell Streak), combining the best features of both device types.[6] Its potential as a games console has also been pointed out.[7]

Like most recent smartphones and tablets, the device is constructed in the "slate" format (see Slate phone and Slate tablet). The body is built from plastic with a metallised rim.[8] The front panel is Gorilla Glass,[9] a strengthened glass often used for high-end devices such as this.

The front panel houses one physical "home" button (for activating the device and switching to the home screen), two illuminated touch pads ("menu" and "back"), the display, the front-facing camera, and light and proximity sensors. At the back is a thin plastic snap-on panel with an indent for a fingernail to facilitate removal, for access to the 2500mAh battery, SIM card and SD Card. The back panel houses the speaker and main camera and flash. The metallic rim houses several controls; at the top edge is the 4-pole 3.5mm jack socket for connecting the headset (which incorporates in-canal earphones, FM radio aerial, microphone and volume control) and a pinhole microphone; at the bottom is the micro-USB socket for charging and data transfer, another pinhole microphone, and the well for storing the stylus; and the sides house an on-off button and a volume control.

Hardware and software features

Hardware specifications of the device include:[10]

- a dual-core 1.4Ghz processor, But Sk telecom, KT, LG U+ use 1.5Ghz Snapdragon S3 Chipset
- a 5.3" WXGA (1280 x 800) HD Super AMOLED screen
- an 8-megapixel main camera that can record 1080p video, and a 2-megapixel front-facing camera
- 802.11 b/g/n support for Wi-Fi
- HSPA+ 21Mbps 850/900/1900/2100 mobile network support.
- positioning using both the a-GPS and GLONASS systems.

Stylus

An unusual feature for such a device is a built-in stylus, which Samsung calls the "S Pen".The stylus tucks into the bottom panel of the phone, and can be used in a variety of apps. It can simply replace the use of a finger in situations where precision is needed, but it is also equipped with a "shift" button, which when pressed enables other functions such as grabbing screenshots (which can then be drawn on using the stylus) or starting the S-Note note-taking application, which is a facility for recording sticky-note-type notes which can include drawing/handwriting (using the stylus), text input, and pictures. The stylus is partnered by a Wacom digitiser system, which results in accurate pressure-sensitive input.[11]

Samsung released in late November 2011 an SDK (software development kit) for the stylus so developers can write third-party apps that use it for input.[12] Android 4.0 "Ice Cream Sandwich", which is expected to be made available for the Galaxy Note in 2012, also includes support for stylus input.[13]

The practice of using a stylus with a handheld computer was popular during the era of Personal Digital Assistant (PDA) devices in the 1990s, whose screens lacked the sensitivity of modern touchscreens and so used the stylus as a primary input method. With the advent of touch-controlled smartphones, styluses became unfashionable; Steve Jobs of Apple, which pioneered tablet and smartphone design, famously remarked "if you see a stylus ... they blew it".[14] Reviewers have therefore expressed interest in the rebirth of stylus input. The Galaxy Note implementation has been described as high quality,[15] superior to other recent efforts such as on the HTC Flyer tablet.[11]

Software

The Galaxy Note is equipped with Android 2.3 "Gingerbread", but Samsung have said that they will be releasing an upgrade to Android 4.0 "Ice Cream Sandwich" during 2012.[16] The standard Android user interface is overlaid with Samsung's TouchWiz 4.0 interface, which includes comprehensive support for the device's stylus, among other things.

As with all smartphones, several application programs are pre-loaded on the device. These include the standard Android applications such as email, web browsing, and media playback, as well as some programs aimed mainly at business users, such as Polaris Office, personal information manager software, the note-taking application, and one game, Crayon Physics Deluxe.

Text input

The Galaxy Note features a variety of text-input methods. A conventional on-screen qwerty keyboard is provided, and there is also the option of handwriting recognition using the stylus, and the time-saving Swype keyboard input method, which replaces direct typing of a word with a single movement of the finger across the desired keys.[17]

Cameras

The device features an 8-megapixel (2448 x 3264 pixels) main camera with flash on the back panel, and a 2-megapixel (1200 x 1600 pixels) camera on the front, primarily for video phone calls. The clear lens cover of the main camera lies flush with the back panel, exposing it to the same scratches that the panel might suffer, which has been a source of criticism.[5] However, the quality of images captured with the main camera has been praised.[5] Video recording through the main camera is 1080p ("Full HD") at 30 frames per second. Photo-editing and video-editing software is supplied with the device.

Variants and customisation

The Galaxy Note was initially produced with a black body (dubbed "carbon blue" by Samsung), but a white body was later also made available.[18]

The following performance variants of the Galaxy Note have been made available:

- N7000 - The original version, with a dual-core 1.4GHz Cortex A9 processor
- N7003 - A lower-powered and cheaper version, with a single-core Qualcomm Snapdragon 8255T processor with an S-LCD screen initially available in South Africa.[19]
- LTE version - with higher-speed communications ability, initially available only in South Korea[20]
- An AT&T version for the U.S. market, featuring LTE, a faster 1.5GHz processor, and a redesigned front panel with four soft buttons instead of one hard and two soft.[3]

Samsung have made available a collection of accessories such as a clip-on screen cover (which replaces the back panel), a docking station, and spare chargers and styluses.[1]

References

[1] "Samsung announces Galaxy Nexus and Note roll-out schedules" (http://www.gsmarena.com/samsung_announces_galaxy_note_and_nexus_roll_out_schedules-news-3322.php). GSMArena. 2011-10-27. .

[2] Mat Smith (2011-12-29). "1 million Galaxy Notes shipped worldwide, US fans throw money at their screens" (http://www.engadget.com/2011/12/29/1-million-galaxy-notes-shipped-worldwide-us-fans-throw-money-at/). Engadget. . Retrieved 2011-12-30.

[3] Myriam Joire (2012-01-09). "AT&T Galaxy Note with LTE hands-on at CES 2012 (video)" (http://www.engadget.com/2012/01/09/atandt-galaxy-note-with-lte-hands-on-at-ces-2012-video/). Engadget. . Retrieved 2012-01-19.

[4] "Samsung Galaxy Note Android phone" (http://www.pcworld.idg.com.au/review/mobile_phones/samsung/galaxy_note/401248). PC World Australia. . Retrieved 2011-12-04.

[5] Dan Grabham (2011-09-01). "Hands on: Samsung Galaxy Note review" (http://www.techradar.com/news/mobile-computing/tablets/hands-on-samsung-galaxy-note-review-1008827). techradar.com. . Retrieved 2011-09-27.

[6] Kevin C. Tofel (2011-11-10). "Can Samsung's 5.3" Galaxy Note bridge phone and tablet use?" (http://gigaom.com/mobile/samsung-galaxy-note-impressions-phone-tablet/). GigaOM. . Retrieved 2011-12-04.

[7] Rik Henderson (2011-11-03). "Samsung Galaxy Note review" (http://www.pocket-lint.com/review/5615/samsung-galaxy-note-review). pocket-lint.com. . Retrieved 2011-12-04.

[8] "Samsung Galaxy Note N7000 review: Power play" (http://www.gsmarena.com/samsung_galaxy_note_n7000-review-676.php). GSMArena. 2011-11-16. . Retrieved 2011-12-11.

[9] "Corning Gorilla Glass - Full Product List" (http://www.corninggorillaglass.com/products-with-gorilla/full-products-list). Corning. . Retrieved 2012-01-20.

[10] Samsung. "Samsung Galaxy Note: Specifications" (http://www.samsung.com/global/microsite/galaxynote/note/spec.html?type=find). . Retrieved 2011-12-05.

[11] Michael Crider (2011-10-27). "Samsung highlights the Galaxy Note's Wacom digitizer" (http://androidcommunity.com/samsung-highlights-the-galaxy-notes-wacom-digitizer-20111027/). androidcommunity.com. . Retrieved 2011-12-04.

[12] Michael Crider (2011-11-28). "Samsung releases S-Pen SDK for the Galaxy Note" (http://androidcommunity.com/samsung-releases-s-pen-sdk-for-the-galaxy-note-20111128/). androidcommunity.com. . Retrieved 2011-12-03.

[13] Donald Melanson (2011-10-21). "Android Ice Cream Sandwich includes native stylus support" (http://www.engadget.com/2011/10/21/android-ice-cream-sandwich-includes-native-stylus-support/). Engadget. . Retrieved 2011-11-13.

[14] Nilay Patel (2010-04-08). "Jobs: If you see a stylus or a task manager, 'they blew it'" (http://www.engadget.com/2010/04/08/jobs-if-you-see-a-stylus-or-a-task-manager-they-blew-it/). engadget.com. . Retrieved 2011-12-04.

[15] Steven Norris (2011-09-11). "Samsung Galaxy Note review — bigger is better" (http://gearburn.com/2011/11/samsung-galaxy-note-review-bigger-is-better/). gearburn.com. . Retrieved 2011-12-04.

[16] Cosmin Vasile (2012-12-20). "Samsung Announces Ice Cream Sandwich for Galaxy Note and Galaxy S II Coming in Q1 2012" (http://news.softpedia.com/news/Samsung-Announces-Ice-Cream-Sandwich-for-Galaxy-Note-and-Galaxy-S-II-Coming-in-Q1-2012-241768.shtml). Softpedia. . Retrieved 2011-12-20.

[17] Sasha Muller (2011-09-05). "Samsung Galaxy Note review: first look" (http://www.pcpro.co.uk/blogs/2011/09/05/samsung-galaxy-note-review-first-look/). PC Pro. . Retrieved 2011-12-06.

[18] Chris Davies (2011-10-27). "Samsung's white Galaxy Note revealed" (http://www.slashgear.com/samsungs-white-galaxy-note-revealed-27191481/). SlashGear. .

[19] "Samsung Galaxy Note N7000 Firmware Updates (Kies Official Release) List" (http://www.webcitation.org/63cUgocLt). Android ROMs. Archived from the original (http://devsrom4android.blogspot.com/2011/11/samsung-galaxy-note-n7000-firmware.html) on 1 December 2011. . Retrieved 2 December 2011.

[20] Michael Crider (2011-11-28). "Samsung Galaxy Note LTE gets official in Korea" (http://androidcommunity.com/samsung-galaxy-note-lte-gets-official-in-korea-20111128/). androidcommunity.com. . Retrieved 2011-12-03.

External links

- Official Samsung Galaxy Note website (http://www.samsung.com/global/microsite/galaxynote/note/index.html?type=find)

Android_(operating_system)

Home screen displayed by Samsung Galaxy Nexus, running Android 4.0 "Ice Cream Sandwich"

Company / developer	Google Inc, Open Handset Alliance
Programmed in	C (core),[1] Java (UI), C++
Working state	Current
Source model	Open Source[2] [3]
Initial release	20 September 2008
Latest stable release	4.0.3 (Ice Cream Sandwich) / 16 December 2011[4]
Package manager	Android Market / APK
Supported platforms	ARM, MIPS,[5] x86 [6] [7]
Kernel type	Monolithic (Linux kernel)
Default user interface	Graphical
License	Apache License 2.0 Linux kernel patches under GNU GPL v2[8]
Official website	www.android.com [9]

Android is a Linux-based operating system for mobile devices such as smartphones and tablet computers. It is developed by the Open Handset Alliance led by Google.[10] [11]

Google purchased the initial developer of the software, Android Inc., in 2005.[12] The unveiling of the Android distribution in 2007 was announced with the founding of the Open Handset Alliance, a consortium of 86 hardware, software, and telecommunication companies devoted to advancing open standards for mobile devices.[13] [14] [15] [16] Google releases the Android code as open-source, under the Apache License.[17] The Android Open Source Project (AOSP) is tasked with the maintenance and further development of Android.[18]

Android has a large community of developers writing applications ("apps") that extend the functionality of the devices. Developers write primarily in a customized version of Java.[19] Apps can be downloaded from third-party sites or through online stores such as Android Market, the app store run by Google. As of October 2011 there were more than 400,000 apps available for Android, and the estimated number of applications downloaded from the

Android Market as of December 2011 exceeded 10 billion.[20] [21]

Android was listed as the best-selling smartphone platform worldwide in Q4 2010 by Canalys[22] [23] with over 200 million Android devices in use by November 2011.[24] According to Google's Andy Rubin, as of December 2011 there are over 700,000 Android devices activated every day.[25]

History

Foundation

Android, Inc. was founded in Palo Alto, California, United States in October, 2003 by Andy Rubin (co-founder of Danger),[26] Rich Miner (co-founder of Wildfire Communications, Inc.),[27] Nick Sears (once VP at T-Mobile),[28] and Chris White (headed design and interface development at WebTV)[29] to develop, in Rubin's words "...smarter mobile devices that are more aware of its owner's location and preferences".[30] Despite the obvious past accomplishments of the founders and early employees, Android Inc. operated secretly, revealing only that it was working on software for mobile phones.[30] That same year, Rubin ran out of money. Steve Perlman, a close friend of Rubin, brought him $10,000 in cash in an envelope and refused a stake in the company.[31]

Acquisition by Google

Google acquired Android Inc. on August 17, 2005, making Android Inc. a wholly owned subsidiary of Google Inc. Key employees of Android Inc., including Andy Rubin, Rich Miner and Chris White, stayed at the company after the acquisition.[12] Not much was known about Android Inc. at the time of the acquisition, but many assumed that Google was planning to enter the mobile phone market with this move.[12]

Post-acquisition development

At Google, the team led by Rubin developed a mobile device platform powered by the Linux kernel. Google marketed the platform to handset makers and carriers on the promise of providing a flexible, upgradable system. Google had lined up a series of hardware component and software partners and signaled to carriers that it was open to various degrees of cooperation on their part.[32] [33] [34]

Speculation about Google's intention to enter the mobile communications market continued to build through December 2006.[35] Reports from the BBC and *The Wall Street Journal* noted that Google wanted its search and applications on mobile phones and it was working hard to deliver that. Print and online media outlets soon reported rumors that Google was developing a Google-branded handset. Some speculated that as Google was defining technical specifications, it was showing prototypes to cell phone manufacturers and network operators.

In September 2007, *InformationWeek* covered an Evalueserve study reporting that Google had filed several patent applications in the area of mobile telephony.[36] [37]

Open Handset Alliance

On November 5, 2007, the Open Handset Alliance, a consortium of several companies which include Broadcom Corporation, Google, HTC, Intel, LG, Marvell Technology Group, Motorola, Nvidia, Qualcomm, Samsung Electronics, Sprint Nextel, T-Mobile and Texas Instruments unveiled itself. The goal of the Open Handset Alliance is to develop open standards for mobile devices.[15] On the same day, the Open Handset Alliance also unveiled their first product, Android, a mobile device platform built on the Linux kernel version 2.6.[15]

On December 9, 2008, 14 new members joined, including ARM Holdings, Atheros Communications, Asustek Computer Inc, Garmin Ltd, Huawei Technologies, PacketVideo, Softbank, Sony Ericsson, Toshiba Corp, and Vodafone Group Plc.[38] [39]

Android Open Source Project

The Android Open Source Project (AOSP) [40] is led by Google, and is tasked with the maintenance and development of Android.[41] According to the project "The goal of the Android Open Source Project is to create a successful real-world product that improves the mobile experience for end users."[42] AOSP also maintains the *Android Compatibility Program*, defining an "Android compatible" device "as one that can run any application written by third-party developers using the Android SDK and NDK", to prevent incompatible Android implementations.[42] The compatibility program is also optional and free of charge, with the *Compatibility Test Suite* also free and open-source.[43]

Version history

Android has seen a number of updates since its original release, each fixing bugs and adding new features. Each version is named, in alphabetical order, after a dessert.[44]

Puppy toy by Eero Aarnio at the Googleplex, 2008

Recent releases

- **2.3 Gingerbread** refined the user interface, improved the soft keyboard and copy/paste features, improved gaming performance, added SIP support (VoIP calls), and added support for Near Field Communication.[45]
- **3.0 Honeycomb** was a tablet-oriented[46] [47] [48] release which supports larger screen devices and introduces many new user interface features, support for multi-core processors, hardware acceleration for graphics[49] and full system encryption.[50] [51] The first device featuring this version, the Motorola Xoom tablet, went on sale in February 2011.[52] [53]
  - **3.1 Honeycomb**, released in May 2011, added support for extra input devices, USB host mode for transferring information directly from cameras and other devices, and the Google Movies and Books apps.[54]
  - **3.2 Honeycomb**, released in July 2011, added optimization for a broader range of screen sizes, new "zoom-to-fill" screen compatibility mode, loading media files directly from SD card, and an extended screen support API.[55] Huawei MediaPad is the first 7 inch tablet to use this version [56]
- **4.0 Ice Cream Sandwich**, announced on October 19, 2011, brought Honeycomb features to smartphones and added new features including facial recognition unlock, network data usage monitoring and control, unified social networking contacts, photography enhancements, offline email searching, app folders, and information sharing using NFC. Android 4.0.3 Ice Cream Sandwich is the latest Android version that is available to phones. The source code of Android 4.0.1 was released on November 14, 2011.[57]

Design

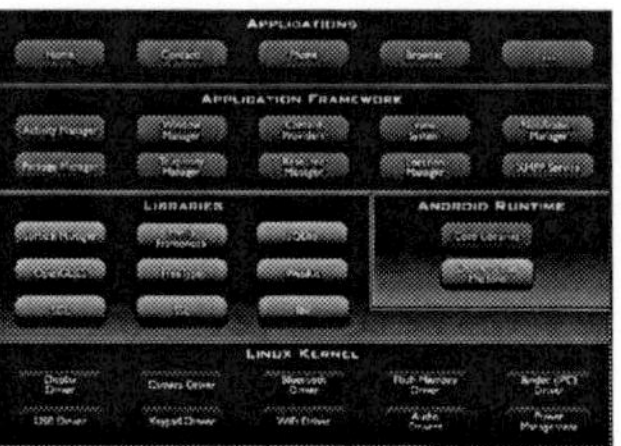
Architecture diagram

Android consists of a kernel based on the Linux kernel, with middleware, libraries and APIs written in C and application software running on an application framework which includes Java-compatible libraries based on Apache Harmony. Android uses the Dalvik virtual machine with just-in-time compilation to run Dalvik dex-code (Dalvik Executable), which is usually translated from Java bytecode.[58]

Linux

Android's kernel is based on the Linux kernel and has further architecture changes by Google outside the typical Linux kernel development cycle.[59] Android does not have a native X Window System nor does it support the full set of standard GNU libraries, and this makes it difficult to port existing Linux applications or libraries to Android.[60]

Certain features that Google contributed back to the Linux kernel, notably a power management feature called wakelocks, were rejected by mainline kernel developers, partly because kernel maintainers felt that Google did not show any intent to maintain their own code.[61] [62] [63] Even though Google announced in April 2010 that they would hire two employees to work with the Linux kernel community,[64] Greg Kroah-Hartman, the current Linux kernel maintainer for the -stable branch, said in December 2010 that he was concerned that Google was no longer trying to get their code changes included in mainstream Linux.[62] Some Google Android developers hinted that "the Android team was getting fed up with the process", because they were a small team and had more urgent work to do on Android.[65]

However, in September 2010, Linux kernel developer Rafael J. Wysocki added a patch that improved the mainline Linux wakeup events framework. He said that Android device drivers that use wakelocks can now be easily merged into mainline Linux, but that Android's opportunistic suspend features should not be included in the mainline kernel.[66] [67] In 2011 Linus Torvalds said that "eventually Android and Linux would come back to a common kernel, but it will probably not be for four to five years".[68]

In December 2011, Greg Kroah-Hartman announced the start of the Android Mainlining Project, which aims to put some Android drivers, patches and features back into the Linux kernel, starting in Linux 3.3.[69]

Features

Current features and specifications:[70] [71] [72]

The Android Emulator default home screen (v1.5)

Handset layouts

The platform is adaptable to larger, VGA, 2D graphics library, 3D graphics library based on OpenGL ES 2.0 specifications, and traditional smartphone layouts.

Storage

SQLite, a lightweight relational database, is used for data storage purposes.

Connectivity

Android supports connectivity technologies including GSM/EDGE, IDEN, CDMA, EV-DO, UMTS, Bluetooth, Wi-Fi, LTE, NFC and WiMAX.

Messaging

SMS and MMS are available forms of messaging, including threaded text messaging and now Android Cloud To Device Messaging (C2DM) is also a part of Android Push Messaging service.

Multiple language support

Android supports multiple languages.[45]

Web browser

The web browser available in Android is based on the open-source WebKit layout engine, coupled with Chrome's V8 JavaScript engine. The browser scores 100/100 on the Acid3 test on Android 4.0.

Java support

While most Android applications are written in Java, there is no Java Virtual Machine in the platform and Java byte code is not executed. Java classes are compiled into Dalvik executables and run on Dalvik, a specialized virtual machine designed specifically for Android and optimized for battery-powered mobile devices with limited memory and CPU. J2ME support can be provided via third-party applications.

Media support

Android supports the following audio/video/still media formats: WebM, H.263, H.264 (in 3GP or MP4 container), MPEG-4 SP, AMR, AMR-WB (in 3GP container), AAC, HE-AAC (in MP4 or 3GP container), MP3, MIDI, Ogg Vorbis, FLAC, WAV, JPEG, PNG, GIF, BMP.[72]

Streaming media support

RTP/RTSP streaming (3GPP PSS, ISMA), HTML progressive download (HTML5 <video> tag). Adobe Flash Streaming (RTMP) and HTTP Dynamic Streaming are supported by the Flash plugin.[73] Apple HTTP Live Streaming is supported by RealPlayer for Android,[74] and by the operating system in Android 3.0 (Honeycomb).[49]

Additional hardware support

Android can use video/still cameras, touchscreens, GPS, accelerometers, gyroscopes, barometers, magnetometers, dedicated gaming controls, proximity and pressure sensors, thermometers, accelerated 2D bit blits (with hardware orientation, scaling, pixel format conversion) and accelerated 3D graphics.

Multi-touch

Android has native support for multi-touch which was initially made available in handsets such as the HTC Hero. The feature was originally disabled at the kernel level (possibly to avoid infringing Apple's patents on touch-screen technology at the time).[75] Google has since released an update for the Nexus One and the

Motorola Droid which enables multi-touch natively.[76]

Bluetooth

Supports A2DP, AVRCP, sending files (OPP), accessing the phone book (PBAP), voice dialing and sending contacts between phones. Keyboard, mouse and joystick (HID) support is available in Android 3.1+, and in earlier versions through manufacturer customizations and third-party applications.[77]

Video calling

Android does not support native video calling, but some handsets have a customized version of the operating system that supports it, either via the UMTS network (like the Samsung Galaxy S) or over IP. Video calling through Google Talk is available in Android 2.3.4 and later. Gingerbread allows Nexus S to place Internet calls with a SIP account. This allows for enhanced VoIP dialing to other SIP accounts and even phone numbers. Skype 2.1 offers video calling in Android 2.3, including front camera support.

Multitasking

Multitasking of applications is available.[78]

Voice based features

Google search through voice has been available since initial release.[79] Voice actions for calling, texting, navigation, etc. are supported on Android 2.2 onwards.[80]

Tethering

Android supports tethering, which allows a phone to be used as a wireless/wired Wi-Fi hotspot. Before Android 2.2 this was supported by third-party applications or manufacturer customizations.[81]

Screen capture

Android supports capturing a screenshot by pressing the power and volume-down buttons at the same time.[82] Prior to Android 4.0, the only methods of capturing a screenshot were through manufacturer and third-party customizations or otherwise by using a PC connection (DDMS developer's tool). These alternative methods are still available with the latest Android.

Uses

While Google has their own line of Android smartphones, the Google Nexus series, the open and customizable nature of the Android operating system allows it to be used on most electronics, including but not limited to: smartphones, fixed phones,[85] laptops, netbooks, smartbooks,[86] [87] tablet computers, E-book readers,[88] TVs (Google TV), wristwatches,[89] [90] headphones,[91] Car CD and DVD players,[92] digital cameras,[93] [94] [95] [96] Portable media players[97] and other devices.[98]

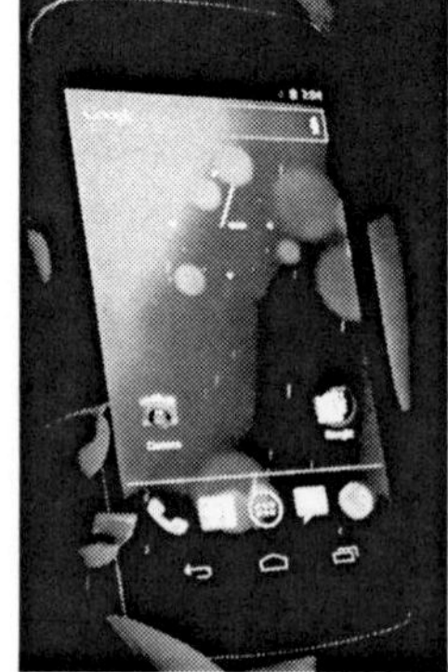

Galaxy Nexus, the latest "Google phone"

The main hardware platform for Android is the ARM architecture. There is support for x86 from the Android x86 project,[99] and Google TV uses a special x86 version of Android.

The first commercially available phone to run Android was the HTC Dream, released on 22 October 2008.[100] In early 2010 Google collaborated with HTC to launch its flagship[101] Android device, the Nexus One. This was followed later in 2010 with the Samsung-made Nexus S and in 2011 with the Galaxy Nexus.

I'm Watch, a wristwatch with phone functionality, running a custom version of Android[83] [84]

Google TV Home Screen

iOS and Android 2.3.3 'Gingerbread' may be set up to dual boot on a jailbroken iPhone or iPod Touch with the help of OpeniBoot and iDroid.[102] [103]

The NOVO7, manufactured by the Chinese company Ainol Electronics, was the world's first Android 4.0 Ice Cream Sandwich tablet.[104]

In December 2011 it was announced the Pentagon has officially approved Android for use by its personnel.[105] [106] [107]

Applications

Applications are usually developed in the Java language using the Android Software Development Kit, but other development tools are available, including a Native Development Kit for applications or extensions in C or C++, Google App Inventor, a visual environment for novice programmers and various cross platform mobile web applications frameworks .

Android Market

The Android Market on a phone

Android Market is the online software store developed by Google for Android devices. An application program ("app") called "Market" is preinstalled on most Android devices and allows users to browse and download apps published by third-party developers, hosted on Android Market. As of October 2011 there were more than 300,000 apps available for Android, and the estimated number of applications downloaded from the Android Market as of December 2011 exceeded 10 billion.[20] [21] The operating system itself is installed on 130 million total devices.[108]

Only devices that comply with Google's compatibility requirements are allowed to preinstall Google's closed-source Android Market app and access the Market.[109] The Market filters the list of applications presented by the Market app to those that are compatible with the user's device, and developers may restrict their applications to particular carriers or countries for business reasons.[110]

Google has participated in the Android Market by offering several applications themselves, including Google Voice (for the Google Voice service), Sky Map (for watching stars), Finance (for their finance service), Maps Editor (for their MyMaps service), Places Directory (for their Local Search), Google Goggles that searches by image, Gesture Search (for using finger-written letters and numbers to search the contents of the phone), Google Translate, Google Shopper, Listen for podcasts and My Tracks, a jogging application. In August

2010, Google launched "Voice Actions for Android",[111] which allows users to search, write messages, and initiate calls by voice.

Alternatively, users can install apps directly onto the device if they have the application's APK file or from third party app stores such as the Amazon Appstore,.[112]

Application security

Android applications run in a sandbox, an isolated area of the operating system that does not have access to the rest of the system's resources, unless access permissions are granted by the user when the application is installed. Before installing an application, Android Market displays all required permissions. A game may need to enable vibration, for example, but should not need to read messages or access the phonebook. After reviewing these permissions, the user can decide whether to install the application.[113]

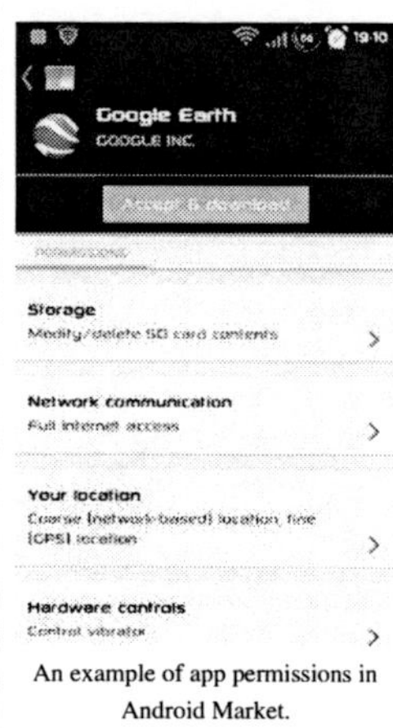

An example of app permissions in Android Market.

Some Android malware incidents have been reported involving rogue applications on Android Market. In August 2010, Kaspersky Lab reported detection of the first malicious program for Android, named Trojan-SMS.AndroidOS.FakePlayer.a, an SMS trojan which had already infected a number of devices.[114] In some cases applications which contained Trojans were hidden in pirated versions of legitimate apps.[115] [116] Google has responded by removing malicious apps from the Android Market, and remotely disabling them on infected devices.[117] Several security firms have released antivirus software for Android devices, in particular, AVG Technologies,[118] Avast!,[119] F-Secure,[120] Kaspersky,[121] McAfee[122] and Symantec.[123]

Privacy

Android smartphones have the ability to report the location of Wi-Fi access points, encountered as phone users move around, to build vast databases containing the physical locations of hundreds of millions of such access points. These databases form electronic maps to locate smartphones, allowing them to run apps like Foursquare, Latitude, Places, and to deliver location-based ads.[124]

Third party monitoring software such as TaintDroid,[125] an academic research-funded project, can, in some cases, detect when personal information is being sent from applications to remote servers.[126]

Marketing

The Android logo was designed along with the Droid font family made by Ascender Corporation.[127]

Android Green is the color of the Android Robot that represents the Android operating system. The print color is PMS 376C and the RGB color value in hexadecimal is #A4C639, as specified by the Android Brand Guidelines.[128] The custom typeface of Android is called Norad (cf. NORAD). It is only used in the text logo.[129]

Market share

Research company Canalys estimated in Q2 2009 that Android had a 2.8% share of worldwide smartphone shipments.[130] By Q4 2010 this had grown to 33% of the market, becoming the top-selling smartphone platform. This estimate includes the Tapas and OMS variants of Android.[22] By Q3 2011 Gartner estimates more than half (52.5%) of the smartphone market belongs to Android.[131]

In February 2010 ComScore said the Android platform had 9.0% of the U.S. smartphone market, as measured by current mobile subscribers. This figure was up from an earlier estimate of 5.2% in November 2009.[132] By the end of Q3 2010 Android's U.S. market share had grown to 21.4%.[133]

In May 2010, Android's first quarter U.S. sales surpassed that of the rival iPhone platform. According to a report by the NPD group, Android achieved 25% smartphone sales in the US market, up 8% from the December quarter. In the second quarter, Apple's iOS was up by 11%, indicating that Android is taking market share mainly from RIM, and still has to compete with heavy consumer demand for new competitor offerings.[134] Furthermore, analysts pointed to advantages that Android has as a multi-channel, multi-carrier OS.[135] In Q4 2010 Android had 59% of the total installed user base of Apple's iOS in the U.S. and 46% of the total installed user base of iOS in Europe.[136] [137]

As of June 2011, Google said that 550,000 new Android devices were being activated every day[138] — up from 400,000 per day a month earlier — and more than 100 million devices had been activated.[139] Android hit 300,000 activations per day back in December 2010. By July 14, 2011, 550,000 Android devices were being activated by Google each day, with 4.4% growth per week.[140] On the 1st of August 2011, Canalys estimated that Android had about 48% of the smartphone market share.[141] On October 13, 2011, Google announced that there were 190 million Android devices in the market.[142] As of November 16, 2011, during the Google Music announcement "These Go to Eleven", 200 million Android devices had been activated.[143] Based on this number, with 1.9% of Android devices being tablets, approximately 3.8 million Android Honeycomb Tablets have been sold.[144] On December 20, 2011. Andy Rubin announced that Google was activating 700,000 new Android devices daily.[25]

Usage share

Usage share of the different versions, by January 3, 2012.[145]

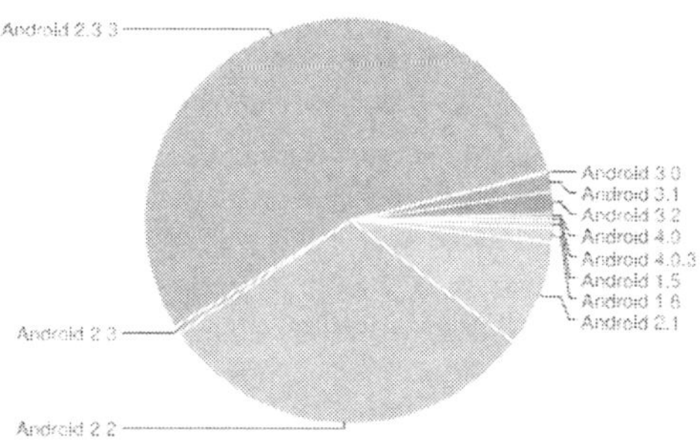

Usage share of the different versions, by January 3, 2012

Distribution	API level	%
4.0.x *Ice Cream Sandwich*	14-16	0.6%
3.x.x *Honeycomb*	11-13	3.3%
2.3.x *Gingerbread*	9-10	55.5%
2.2 *Froyo*	8	30.4%
2.0, 2.1 *Eclair*	7	8.5%
1.6 *Donut*	4	1.1%
1.5 *Cupcake*	3	0.6%

There were two more internal releases, called "Astro" and "Bender". The code names are in alphabetical order, and were allegedly changed from robots to desserts to avoid trademark issues.[146]

Retail stores

The carrier Telstra opened the world's first Android store, Androidland, on Bourke Street, Melbourne in December 2011.[147]

Intellectual property

Trademarks

In order to use the Android trademark, device manufacturers must ensure that the device complies with the Compatibility Definition Document (CDD) and then get permission from Google. Devices must also meet this definition to be eligible to license Google's closed-source applications, including the Android Market.[148] Participation in the compatibility program is free of charge.[43]

In September 2010, Skyhook Wireless filed a lawsuit against Google in which they alleged that Google had used the compatibility document to block Skyhook's mobile positioning service (XPS) from Motorola's Android mobile devices.[149] In December 2010 a judge denied Skyhook's motion for preliminary injunction, saying that Google had not closed off the possibility of accepting a revised version of Skyhook's XPS service, and that Motorola had terminated their contract with Skyhook because Skyhook wanted to disable Google's location data collection functions on Motorola's devices, which would have violated Motorola's obligations to Google and its carriers.[150]

Licensing

The source code for Android is available under free and open source software licenses. Google published their Linux kernel changes under the GNU General Public License version 2, and the rest of the code (including network and telephony stacks)[151] under the Apache License version 2.0.[152] [153] [154] Google also keeps the reviewed issues list publicly open for anyone to see and comment.[155]

The Open Handset Alliance develops the GPL-licensed part of Android, that is their changes to the Linux kernel, in public, with source code publicly available at all times. The rest of Android is developed in private, with source code released publicly when a major new version is released. Typically Google collaborates with a hardware manufacturer to produce a flagship device (part of the Google Nexus series) featuring the new version of Android, then makes the source code available after that device has been released.[156]

In early 2011, Google chose to temporarily withhold the Android source code to the tablet-only Honeycomb release, creating doubts over Google's commitment to open source with Android.[157] The reason, according to Andy Rubin in an official Android blog post, was because Honeycomb was rushed for production of the Motorola Xoom,[158] and they did not want third parties creating a "really bad user experience" by attempting to put onto smartphones a version of Android intended for tablets.[159] The source code was once again made available in November 2011 with the release of Android 4.0.[160]

Patents

Both Android and Android phone manufacturers have been the target of numerous patent lawsuits. On 12 August 2010, Oracle sued Google over claimed infringement of copyrights and patents related to the Java programming language.[161] Specifically, the patent infringement claim references seven United States patents including US 5966702 [162], and US 6910205 [163].[164]

In response, Google submitted multiple lines of defense, counterclaiming that Android did not infringe on Oracle's patents or copyright, that Oracle's patents were invalid, and several other defenses. They said that Android is based on Apache Harmony, a clean room implementation of the Java class libraries, and an independently developed virtual machine called Dalvik.[165] [166] [167]

Microsoft has also sued several manufacturers of Android devices for patent infringement, and collects patent licensing fees from others. In October 2011 Microsoft said they had signed license agreements with ten Android device manufacturers, accounting for 55% of worldwide revenue for Android devices.[168] These include Samsung and HTC.[169]

Google has publicly expressed its dislike for the current patent landscape in the United States, accusing Apple, Oracle and Microsoft of trying to take down Android through patent litigation, rather than innovating and competing with better products and services.[170] In August 2011, Google started the process of purchasing Motorola Mobility for US$12.5 billion, which was viewed in part as a defensive measure to protect Android, since Motorola Mobility holds more than 17,000 patents.[171] In December 2011 Google acquired in the region of a thousand patents from IBM,[172] which may aid in defense against Oracle.[173]

See also

- CyanogenMod
- Google Android lawn statues
- Google Chrome OS
- Google Nexus
- List of Android devices
- Index of Android OS articles
- Rooting (Android OS)

References

[1] Lextrait, Vincent (January 2010). "The Programming Languages Beacon, v10.0" (http://www.lextrait.com/Vincent/implementations.html). . Retrieved 2010-01-05.
[2] "Philosophy and Goals | Android Open Source" (http://source.android.com/about/philosophy.html). Source.android.com. . Retrieved 2012-01-01.
[3] "Codenames, Tags, and Build Numbers | Android Open Source" (http://source.android.com/source/build-numbers.html). Source.android.com. 2010-06-24. . Retrieved 2011-08-08.
[4] Xavier Ducrohet (16 December 2011). "Android 4.0.3 Platform and Updated SDK tools" (http://android-developers.blogspot.com/2011/12/android-403-platform-and-updated-sdk.html). *Android Developers Blog*. . Retrieved 17 December 2011.
[5] "MIPS gets sweet with Honeycomb" (http://www.eetimes.com/electronics-news/4215490/MIPS-gets-sweet-with-Honeycomb). Eetimes.com. . Retrieved 2011-08-08.
[6] "Porting Android to x86" (http://www.android-x86.org/). Android-x86. . Retrieved 2011-08-08.
[7] Agam Shah (2011-12-01). "Google's Android 4.0 ported to x86 processors" (http://www.computerworld.com/s/article/9222323/Google_s_Android_4.0_ported_to_x86_processors). Computerworld. . Retrieved 2011-12-19.
[8] "Licenses" (http://source.android.com/source/licenses.html). *Android Open Source Project*. Open Handset Alliance. . Retrieved 2010-06-10.
[9] http://www.android.com/
[10] "Google Projects for Android" (http://www.webcitation.org/5wiw1JXa2). *code.google.com*. Google Inc. 2011. Archived from the original (http://code.google.com/android/) on 2011-02-23. . Retrieved 2011-02-23.
[11] "Philosophy and Goals" (http://www.webcitation.org/5wiy036ap). *source.android.com*. Google Inc. 2011. Archived from the original (http://source.android.com/about/philosophy.html) on 2011-02-23. . Retrieved 2011-02-23.
[12] "Google Buys Android for Its Mobile Arsenal" (http://www.businessweek.com/technology/content/aug2005/tc20050817_0949_tc024.htm). Businessweek.com. 2005-08-17. . Retrieved 2010-10-29.
[13] "Open Handset Alliance" (http://www.openhandsetalliance.com/). Open Handset Alliance. . Retrieved 2010-06-10.
[14] Jackson, Rob (10 December 2008). "Sony Ericsson, HTC Androids Set For Summer 2009" (http://phandroid.com/2008/12/10/sony-ericsson-htc-androids-set-for-summer-2009/). *Android Phone Fans*. . Retrieved 2009-09-03.
[15] "Industry Leaders Announce Open Platform for Mobile Devices" (http://www.openhandsetalliance.com/press_110507.html) (Press release). Open Handset Alliance. 2007-11-05. . Retrieved 2007-11-05.
[16] "FAQ" (http://www.openhandsetalliance.com/oha_faq.html) (Press release). Open Handset Alliance. . Retrieved 2010-11-15.
[17] "Android Overview" (http://www.openhandsetalliance.com/android_overview.html). Open Handset Alliance. . Retrieved 2008-09-23.
[18] "About the Android Open Source Project" (http://source.android.com/about/index.html). . Retrieved 2010-11-15.

[19] Shankland, Stephen (12 November 2007). "Google's Android parts ways with Java industry group" (http://www.news.com/8301-13580_3-9815495-39.html). *CNET News*. .

[20] "Android Market reaches 500,000 app mark" (http://www.t3.com/news/android-market-reaches-500000-app-mark). www.t3.com. 2011-10-23. . Retrieved 2011-10-23.

[21] Christina Bonnington (8 December 2011). "Google's 10 Billion Android App Downloads: By the Numbers" (http://www.wired.com/gadgetlab/2011/12/10-billion-apps-detailed/). *wired.com*. . Retrieved 12 December 2011.

[22] Tarmo Virki and Sinead Carew (2011-01-31). "Google topples Symbian from smartphones top spot" (http://uk.reuters.com/article/2011/01/31/oukin-uk-google-nokia-idUKTRE70U1YT20110131). *Reuters*. . Retrieved 1 February 2011.

[23] "Google's Android becomes the world's leading smart phone platform (Canalys research release: r2011013)" (http://www.canalys.com/pr/2011/r2011013.html). *Canalys*. 31 January 2011. . Retrieved 1 February 2011.

[24] Charles Arthur (2011-10-14). "Mobile generating equivalent of $2.5bn a year, says Google chief" (http://www.guardian.co.uk/technology/2011/oct/14/android-google-ad-revenue). Guardian. . Retrieved 2011-10-15.

[25] "Android Phones Pass 700,000 Activations Per Day, Approaching 250 Million Total" (http://techcrunch.com/2011/12/22/android-700000/). TechCrunch. 22 December 2011. . Retrieved 22 December 2011.

[26] Markoff, John (2007-11-04). "I, Robot: The Man Behind the Google Phone" (http://www.nytimes.com/2007/11/04/technology/04google.html?_r=2&hp=&pagewanted=all). The New York Times. . Retrieved 2008-10-14.

[27] Kirsner, Scott (2007-09-02). "Introducing the Google Phone" (http://web.archive.org/web/20100104054533/http://www.boston.com/business/technology/articles/2007/09/02/introducing_the_google_phone/). *The Boston Globe*. Archived from the original (http://www.boston.com/business/technology/articles/2007/09/02/introducing_the_google_phone/) on January 4, 2010. . Retrieved 2008-10-24.

[28] "T-Mobile Brings Unlimited Multiplayer Gaming to US Market with First Launch of Nokia N-Gage Game Deck" (http://www.nokia.com/A4136002?newsid=918410) (Press release). Nokia. 23 September 2003. . Retrieved 2009-04-05.

[29] Elgin, Ben (17 August 2005). "Google Buys Android for Its Mobile Arsenal" (http://www.businessweek.com/technology/content/aug2005/tc20050817_0949_tc024.htm). *BusinessWeek*. . Retrieved 2009-04-23.

[30] Elgin, Ben (2005-08-17). "Google Buys Android for Its Mobile Arsenal" (http://www.webcitation.org/5wk7sIvVb). *businessweek.com*. Bloomberg L.P. Archived from the original (http://www.businessweek.com/technology/content/aug2005/tc20050817_0949_tc024.htm) on 2011-02-24. . Retrieved 2011-02-24. "In what could be a key move in its nascent wireless strategy, Google (GOOG) has quietly acquired startup Android Inc...."

[31] Vance, Ashlee (2011-08-07). "A Thousand Times Yes" (http://www.airsla.org/broadcasts/BusinessWeek110802.mp3). Bloomberg BusinessWeek. . Retrieved 2011-11-09.

[32] Block, Ryan (2007-08-28). "Google is working on a mobile OS, and it's due out shortly" (http://www.engadget.com/2007/08/28/google-is-working-on-a-mobile-os-and-its-due-out-shortly/). *Engadget*. . Retrieved 2007-11-06.

[33] Sharma, Amol; Delaney, Kevin J. (2007-08-02). "Google Pushes Tailored Phones To Win Lucrative Ad Market" (http://online.wsj.com/article_email/SB118602176520985718-lMyQjAxMDE3ODA2MjAwMjIxWj.html). *The Wall Street Journal*. . Retrieved 2007-11-06.

[34] "Google admits to mobile phone plan" (http://www.directtraffic.org/OnlineNews/Google_admits_to_mobile_phone_plan_18094880.html). *directtraffic.org*. Google News. 2007-03-20. . Retrieved 2007-11-06.

[35] McKay, Martha (21 December 2006). "Can iPhone become your phone?; Linksys introduces versatile line for cordless service". *The Record (Bergen County)*: p. L9. "And don't hold your breath, but the same cell phone-obsessed tech watchers say it won't be long before Google jumps headfirst into the phone biz. Phone, anyone?"

[36] Claburn, Thomas (2007-09-19). "Google's Secret Patent Portfolio Predicts gPhone" (http://www.informationweek.com/news/showArticle.jhtml?articleID=201807587&cid=nl_IWK_daily). *InformationWeek*. . Retrieved 2007-11-06.

[37] Pearce, James Quintana (2007-09-20). "Google's Strong Mobile-Related Patent Portfolio" (http://www.moconews.net/entry/419-googles-strong-mobile-related-patent-portfolio/). *mocoNews.net*. . Retrieved 2007-11-07.

[38] Martinez, Jennifer (2008-12-10). "Corrected: Update 2: More mobile phone makers back Google's Android" (http://www.reuters.com/article/newsOne/idUSN0928595620081210). *Reuters* (Thomson Reuters). . Retrieved 2008-12-13.

[39] Kharif, Olga (2008-12-09). "Google's Android Gains More Powerful Followers" (http://www.businessweek.com/the_thread/techbeat/archives/2008/12/googles_android_2.html). *BusinessWeek*. McGraw-Hill. . Retrieved 2008-12-13.

[40] http://source.android.com/

[41] "About the Android Open Source Project | Android Open Source" (http://source.android.com/about/index.html). Source.android.com. . Retrieved 2011-12-29.

[42] "Philosophy and Goals | Android Open Source" (http://source.android.com/about/philosophy.html). Source.android.com. . Retrieved 2011-12-29.

[43] "Frequently Asked Questions | Android Open Source" (http://source.android.com/faqs.html#compatibility). Source.android.com. . Retrieved 2011-12-29.

[44] John D. Sutter (4 February 2011). "Why does Google name its Android products after desserts?" (http://articles.cnn.com/2011-02-04/tech/google.honeycomb.android.names_1_google-android-android-os-randall-sarafa?_s=PM:TECH). *CNNTech*. . Retrieved 21 October 2011.

[45] "Android 2.3 Platform Highlights" (http://developer.android.com/sdk/android-2.3-highlights.html). *Android Developers*. 6 December 2010. . Retrieved 2010-12-07.

[46] Mithun Chandrasekhar (2 February 2011). "Google's Android Event Analysis" (http://www.anandtech.com/show/4150/googles-android-event-analysis/2). *AnandTech*. . Retrieved 5 February 2011. "I confirmed this with Google; Honeycomb, at least in the current form, will not be coming to non-tablet devices."

[47] Rapheal, JR. "Will Android Honeycomb come to smartphones?" (http://blogs.computerworld.com/17642/android_honeycomb_smartphones). *Computerworld*. . Retrieved 24 February 2011.

[48] "Android Platform Highlights" (http://developer.android.com/sdk/android-3.0-highlights.html). *Google*. . Retrieved 24 February 2011.

[49] "Android 3.0 Platform Highlights" (http://developer.android.com/sdk/android-3.0-highlights.html). *Android Developers*. 26 January 2011. . Retrieved 2011-01-26.

[50] "Notes on the implementation of encryption in Android 3.0" (http://source.android.com/tech/encryption/android_crypto_implementation.html). *Android.com*. . Retrieved 29 December 2011. "If you want to enable encryption on your device based on Android 3.0 aka Honeycomb"

[51] "Encrypt the phone's internal storage" (http://android.stackexchange.com/questions/4567/encrypt-the-phones-internal-storage). *Android Enthusiasts - Stack Exchange*. 8 May 2011. . Retrieved 29 December 2011. "Android 3 (Honeycomb) offers full system encryption natively."

[52] Nilay Patel (26 January 2011). "Motorola Atrix 4G and Xoom tablet launching at the end of February, Droid Bionic and LTE Xoom in Q2" (http://www.engadget.com/2011/01/26/motorola-atrix-4g-and-xoom-tablet-launching-at-the-end-of-februa/). *Engadget*. . Retrieved 5 February 2011.

[53] German, Kent (2011-10-18). "Ice Cream Sandwich adds tons of new features" (http://reviews.cnet.com/8301-19736_7-20122331-251/ice-cream-sandwich-adds-tons-of-new-features/?tag=mncol;txt). Reviews.cnet.com. . Retrieved 2011-11-18.

[54] Donald Melenson (10 May 2011). "Google announces Android 3.1, available on Verizon Xoom today" (http://www.engadget.com/2011/05/10/google-announces-android-3-1/). *Engadget*. . Retrieved 23 October 2011.

[55] Crothers, Brooke (2011-07-17). "Android 3.2 official, coming to a tablet near you" (http://news.cnet.com/8301-13924_3-20080221-64/android-3.2-official-coming-to-a-tablet-near-you/). News.cnet.com. . Retrieved 2011-08-08.

[56] Darren Murph (2011-06-20). "Huawei MediaPad revealed first 3.2 tablet" (http://www.engadget.com/2011/06/20/huawei-mediapad-revealed-worlds-first-7-inch-android-3-2-table/). *Engadget*. . Retrieved 2011-06-20.

[57] Brad Molen (18 October 2011). "Android 4.0 Ice Cream Sandwich now official, includes revamped design, enhancements galore" (http://www.engadget.com/2011/10/18/android-4-0-ice-cream-sandwich-now-official/). *Engadget*. . Retrieved 21 October 2011.

[58] Tim Bray (24 November 2010). "What Android Is" (http://www.tbray.org/ongoing/When/201x/2010/11/14/What-Android-Is). *ongoing by Tim Bray*. . Retrieved 27 October 2011.

[59] *Androidology – Part 1 of 3 – Architecture Overview* (http://www.youtube.com/watch?v=QBGfUs9mQYY) (Video). YouTube. 2008-09-06. . Retrieved 2007-11-07.

[60] Paul, Ryan (23 February 2009). "Dream(sheep++): A developer's introduction to Google Android" (http://arstechnica.com/open-source/reviews/2009/02/an-introduction-to-google-android-for-developers.ars). *Ars Technica*. . Retrieved 2009-03-07.

[61] "Linux developer explains Android kernel code removal" (http://news.zdnet.com/2100-9595_22-389733.html). ZDNet. 2010-02-02. . Retrieved 2010-02-03.

[62] Greg Kroah-Hartman (2010-02-02). "Android and the Linux kernel community" (http://www.kroah.com/log/linux/android-kernel-problems.html). . Retrieved 2010-02-03. "*Google shows no sign of working to get their code upstream anymore. Some companies are trying to strip the Android-specific interfaces from their codebase and push that upstream, but that causes a much larger engineering effort, and is a pain that just should not be necessary.*"

[63] "Garrett's LinuxCon Talk Emphasizes Lessons Learned from Android/Kernel Saga" (http://www.linux.com/news/embedded-mobile/mobile-linux/344486-garretta-linuxcon-talk-emphasizes-lessons-learned-from-androidkernel-saga). Linux.com. 2011-08-10. . Retrieved 2011-01-02.

[64] "DiBona: Google will hire two Android coders to work with kernel.org" (http://blogs.zdnet.com/open-source/?p=6274). *www.zdnet.com*. 15 April 2010. . Retrieved 2010-04-29.

[65] "Android/Linux kernel fight continues" (http://blogs.computerworld.com/16900/android_linux_kernel_fight_continues). Computerworld. 2010-09-07. . Retrieved 2011-01-02.

[66] Rafael J. Wysocki (24 November 2010). "An alternative to suspend blockers" (http://lwn.net/Articles/416690/). *lwn.net*. . Retrieved 7 September 2011.

[67] Rafael J. Wysocki (12 November 2010). "Technical Background of the Android Suspend Blockers Controversy" (http://lwn.net/images/pdf/suspend_blockers.pdf). . Retrieved 7 September 2011. "...the most controversial parts of the Android's opportunistic suspend infrastructure are not really necessary and therefore they should not be included into the mainline kernel.... it should be possible to convert the vast majority of the Android device drivers using wakelocks to the mainline kernel code base."

[68] Steven J. Vaughan-Nichols (18 August 2011). "Linus Torvalds on Android, the Linux fork" (http://www.zdnet.com/blog/open-source/linus-torvalds-on-android-the-linux-fork/9426). *zdnet.com*. . Retrieved 8 September 2011.

[69] Chris von Eitzen (23 December 2011). "Android drivers to be included in Linux 3.3 kernel" (http://www.h-online.com/open/news/item/Android-drivers-to-be-included-in-Linux-3-3-kernel-1400996.html). *h-online.com*. . Retrieved 23 December 2011.

[70] "What is Android?" (http://developer.android.com/guide/basics/what-is-android.html). *Android Developers*. 21 July 2009. . Retrieved 2009-09-03.

[71] Topolsky, Joshua (2007-11-12). "Google's Android OS early look SDK now available" (http://www.engadget.com/2007/11/12/googles-android-os-early-look-sdk-now-available/). *Engadget*. . Retrieved 2007-11-12.

[72] "Android Supported Media Formats" (http://developer.android.com/guide/appendix/media-formats.html). *Android Developers*. . Retrieved 2009-05-01.

[73] "Flash Flayer 10.1 for Android 2.2 Release Notes" (http://kb2.adobe.com/cps/860/cpsid_86018.html). *Adobe Knowledgebase*. . Retrieved 27 January 2011.

[74] "RealNetworks Gives Handset and Tablet OEMs Ability to Deliver HTTP Live Content to Android Users" (http://www.realnetworks.com/pressroom/releases/2010/RealPlayer-for-Mobile-Delivers-HTTP-Live-Content-to-Android.aspx). *realnetworks.com*. 10 September 2010. . Retrieved 27 January 2011.

[75] Musil, Steven (11 February 2009). "Report: Apple nixed Android's multitouch" (http://news.cnet.com/8301-13579_3-10161312-37.html). *CNET News*. . Retrieved 2009-09-03.

[76] Ziegler, Chris (2 February 2010). "Nexus One gets a software update, enables multitouch" (http://www.engadget.com/2010/02/02/nexus-one-gets-a-software-update-enables-multitouch/). *Engadget*. . Retrieved 2010-02-02.

[77] "Android 3.1 Platform Highlights" (http://developer.android.com/sdk/android-3.1-highlights.html#UserFeatures). *Android Developers*. . Retrieved 26 August 2011.

[78] Bray, Tim (28 April 2010). "Multitasking the Android Way" (http://android-developers.blogspot.com/2010/04/multitasking-android-way.html). *Android Developers*. . Retrieved 2010-11-03.

[79] "Speech Input for Google Search" (http://developer.android.com/resources/articles/speech-input.html). *Android Developers*. . Retrieved 3 November 2010.

[80] "Voice Actions for Android" (http://www.google.com/mobile/voice-actions/). *google.com*. . Retrieved 27 January 2011.

[81] JR Raphael (6 May 2010). "Use Your Android Phone as a Wireless Modem" (http://www.pcworld.com/article/190265/use_your_android_phone_as_a_wireless_modem.html). PCWorld. . Retrieved 3 November 2010.

[82] Nancy Gohring (19 October 2011). "Samsung, Google Unveil Latest Android OS, Phone" (http://www.pcworld.com/article/242128/samsung_google_unveil_latest_android_os_phone.html). PCWorld. . Retrieved 19 October 2011.

[83] December 29, 2011 by Lauren Hockenson 31. "The Android-Powered Smart Watch Marries Luxury and Tech" (http://mashable.com/2011/12/29/im-watch-design/). Mashable.com. . Retrieved 2012-01-01.

[84] Rob Waugh (2011-11-29). "Time for a chat? New watch-phone puts Android on your wrist | Mail Online" (http://www.dailymail.co.uk/sciencetech/article-2067701/Time-chat-New-watch-phone-puts-Android-wrist.html). Dailymail.co.uk. . Retrieved 2012-01-01.

[85] http://www.androidcentral.com/archos-smart-home-phone-now-available-get-android-your-landline

[86] "Sharp ISO1 Android Smartbook Headed To Japan" (http://phandroid.com/2010/03/30/sharp-iso1-android-smartbook-headed-to-japan/). Phandroid.com. 2010-03-30. . Retrieved 2012-01-01.

[87] Laura June (6 September 2010). "Toshiba AC100 Android smartbook hits the United Kingdom" (http://www.engadget.com/2010/09/06/toshiba-ac100-android-smartbook-hits-the-united-kingdom/). *Engadget*. . Retrieved 9 June 2011.

[88] Jolie O'Dell (12 May 2011). "Androids Unite: How Ice Cream Sandwich Will End the OS Schism" (http://mashable.com/2011/05/12/ice-cream-sandwich/). *Mashable*. . Retrieved 9 June 2011.

[89] "i'm Watch" (http://live.imwatch.it/). Live.imwatch.it. . Retrieved 2011-11-18.

[90] Hollister, Sean (2011-12-21), *Sony Smart Watch (aka Sony Ericsson LiveView 2) hands-on* (http://www.theverge.com/2012/1/10/2695959/sony-smart-watch-aka-sony-ericsson-liveview-2-hands-on), The Verge, , retrieved 2012-01-10

[91] Rik Myslewski (2011-01-12), *Android-powered touchscreen Wi-Fi headphones* (http://www.theregister.co.uk/2011/01/12/now_audio_admiral_touch/), theregister.co.uk, , retrieved 2012-01-15

[92] "Car Player Android-Car Player Android Manufacturers, Suppliers and Exporters on" (http://www.alibaba.com/showroom/car-player-android.html). Alibaba.com. . Retrieved 2011-11-18.

[93] "Altek Leo, the 14 megapixel Android cameraphone, headed for Europe in 2011" (http://www.engadget.com/2010/10/03/altek-leo-the-14-megapixel-android-cameraphone-headed-for-euro/). Engadget. 2010-10-03. . Retrieved 2012-01-04.

[94] "Android powered 14MP camera phone Altek Leo coming to Europe in Q1 next year" (http://androinica.com/2010/10/android-powered-14mp-camera-phone-altek-leo-coming-to-europe-in-q1-next-year/). Androinica.com. 2010-10-04. . Retrieved 2012-01-04.

[95] "Android Phone + Lumix Camera = Panasonic's LUMIX Phone 101P For Japan" (http://techcrunch.com/2011/09/29/android-phone-lumix-camera-panasonics-lumix-phone-101p-for-japan/). TechCrunch. 2011-09-29. . Retrieved 2012-01-04.

[96] "Panasonic Lumix Phone 101P – Digital Camera Meets Android" (http://phandroid.com/2011/09/29/panasonic-lumix-phone-101p-digital-camera-meets-android/). Phandroid.com. 2011-09-29. . Retrieved 2012-01-04.

[97] *Top Android MP3 Players for 2011* (http://www.androidauthority.com/top-android-mp3-players-for-2011-36523/), Androidauthority.com, 2011-12-01, , retrieved 2012-01-10

[98] "Run Android on Your Netbook or Desktop" (http://www.howtogeek.com/howto/22665/run-android-on-your-netbook-or-desktop/). How-To Geek. . Retrieved 2011-11-18.

[99] "Android-x86 - Porting Android to x86" (http://www.android-x86.org/). .

[100] "T-Mobile Unveils the T-Mobile G1 - the First Phone Powered by Android" (http://www.htc.com/www/press.aspx?id=66338&lang=1033). HTC. . Retrieved 2009-05-19. AT&T's first device to run the Android OS was the Motorola Backflip.

[101] Richard Wray (14 March 2010). "Google forced to delay British launch of Nexus phone" (http://www.guardian.co.uk/technology/2010/mar/14/google-mobile-phone-launch-delay). London: guardian.co.uk. .

[102] David Wang (19 May 2010). "How to Install Android on Your iPhone" (http://www.pcworld.com/article/196595/how_to_install_android_on_your_iphone.html). pcworld.com. .

[103] "Idroidproject.org" (http://www.idroidproject.org/). Idroidproject.org. . Retrieved 2011-08-08.
[104] Brad Molen (5 December 2011). "Ainol launches the NOVO7, the world's first Android 4.0 tablet, for $100 plus shipping" (http://www.engadget.com/2011/12/05/ainol-launches-the-novo7-the-worlds-first-android-4-0-tablet/). Engadget. .
[105] Graziano, Dan (2011-12-28), *Pentagon approves Android device for Department of Defense, Apple still awaits clearance* (http://www.bgr.com/2011/12/28/pentagon-approves-android-device-for-department-of-defense-apple-still-awaits-clearance/), Bgr.com, , retrieved 2012-01-12
[106] *Pentagon OKs Android for DoD Usage* (http://www.itproportal.com/2011/12/26/pentagon-oks-android-dod-usage/), ITProPortal.com, 2011-12-26, , retrieved 2012-01-12
[107] *Android to be Approved for DoD Use Within Weeks* (http://www.military.com/news/article/2011/android-to-be-approved-for-dod-use-within-weeks.html), Military.com, 2011-12-09, , retrieved 2012-01-12
[108] "Google Android now on 130M total devices, with 6B app downloads" (http://techcrunch.com/2011/07/14/google-android-now-on-130-million-total-devices/). TechCrunch. 2011-07-14. . Retrieved 2011-08-08.
[109] "Android Compatibility" (http://source.android.com/compatibility/index.html). *Android Open Source Project*. . Retrieved 31 December 2010.
[110] "Android Compatibility" (http://developer.android.com/guide/practices/compatibility.html). *Android Developers*. . Retrieved 31 December 2010.
[111] "Voice Actions for Android" (http://www.google.com/mobile/voice-actions/index.html). Google.com. . Retrieved 2011-08-08.
[112] Ganapati, Priya (June 11, 2010). "Independent App Stores Take On Google's Android Market" (http://www.wired.com/gadgetlab/2010/06/independent-app-stores-take-on-googles-android-market/). Wired News. . Retrieved 2011-02-02.
[113] "Android Security Overview" (http://source.android.com/tech/security/index.html). *Android Open Source Project*. . Retrieved 23 October 2011.
[114] "First SMS Trojan detected for smartphones running Android" (http://www.kaspersky.com/news?id=207576158). Kaspersky Lab. . Retrieved 2010-10-18.
[115] "The Mother Of All Android Malware Has Arrived" (http://www.androidpolice.com/2011/03/01/the-mother-of-all-android-malware-has-arrived-stolen-apps-released-to-the-market-that-root-your-phone-steal-your-data-and-open-backdoor/). *Android Police*. March 6, 2011. .
[116] Perez, Sarah (2009-02-12). "Android Vulnerability So Dangerous, Owners Warned Not to Use Phone's Web Browser" (http://www.readwriteweb.com/archives/android_vulnerability_so_dangerous_shouldnt_use_web_browser.php). Readwriteweb.com. . Retrieved 2011-08-08.
[117] "Google Responds To Android Malware, Will Fix Infected Devices And 'Remote Kill' Malicious Apps" (http://techcrunch.com/2011/03/05/android-malware-rootkit-google-response/). *Tech Crunch*. March 6, 2011. .
[118] *Antivirus for Android Smartphones* (http://www.avg.com/us-en/antivirus-for-android), AVG, , retrieved 2012-01-15
[119] *Mobile Security* (http://www.avast.com/free-mobile-security), Avast.com, , retrieved 2012-01-15
[120] *Mobile Security - System requirements* (http://www.f-secure.com/en/web/home_global/protection/mobile-security/system-requirements), F-Secure, , retrieved 2012-01-15
[121] *Kaspersky Mobile Security* (http://www.kaspersky.com/mobile_downloads), Kaspersky.com, , retrieved 2012-01-15
[122] *McAfee Mobile Security for Android* (https://www.mcafeemobilesecurity.com/products/android.aspx), Mcafeemobilesecurity.com, , retrieved 2012-01-15
[123] *Mobile Internet Security* (http://us.norton.com/mobile-security/), Us.norton.com, 2007-09-25, , retrieved 2012-01-15
[124] Steve Lohr (8 May 2011). "Skyhook Wireless v. Google Case Yields E-Mail Insight" (http://www.nytimes.com/2011/05/09/technology/09google.html?pagewanted=all). *The New York Times* (NYTC). ISSN 0362-4331. . Retrieved 2 November 2011.
[125] "AppAnalysis.org: Real Time Privacy Monitoring on Smartphones" (http://appanalysis.org/faq.html). . Retrieved 2011-06-01.
[126] Kit Eaton (29 September 2010). "TaintDroid Tracks Leaks of Personal Data to Ad Firms" (http://www.fastcompany.com/1692088/android-apps-leaking-your-personal-data-to-ad-firms-says-taintdroid-app). *FastCompany*. . Retrieved 2 November 2011.
[127] Woyke, Elizabeth (26 September 2008). "Android's Very Own Font" (http://www.forbes.com/2008/09/25/font-android-g1-tech-wire-cx_ew_0926font.html). *Forbes*. .
[128] "Brand Guidelines" (http://www.android.com/branding.html). *Android*. 23 March 2009. . Retrieved 2009-10-30.
[129] "Android Brand Guidelines" (http://www.android.com/branding.html). *Android*. 2009-03-23. . Retrieved 2010-04-10.
[130] "Canalys: iPhone outsold all Windows Mobile phones in Q2 2009" (http://www.appleinsider.com/articles/09/08/21/canalys_iphone_outsold_all_windows_mobile_phones_in_q2_2009.html). *AppleInsider*. 21 August 2009. . Retrieved 2009-09-21.
[131] "Gartner Says Sales of Mobile Devices Grew 5.6 Percent in Third Quarter of 2011; Smartphone Sales Increased 42 Percent" (http://www.gartner.com/it/page.jsp?id=1848514). 15 November 2011. . Retrieved 16 November 2011.
[132] "comScore Reports February 2010 U.S. Mobile Subscriber Market Share" (http://www.mycomscore.net/Press_Events/Press_Releases/2010/4/comScore_Reports_February_2010_U.S._Mobile_Subscriber_Market_Share). *Comscore.com*. 5 April 2010. . Retrieved 24 December 2010. "RIM, 42.1%; Apple, 25.4%; Microsoft, 15.1%; Google (Android), 9.0%; Palm, 5.4%; others, 3.0%"
[133] "comScore Reports September 2010 U.S. Mobile Subscriber Market Share" (http://www.comscore.com/Press_Events/Press_Releases/2010/11/comScore_Reports_September_2010_U.S._Mobile_Subscriber_Market_Share). *Comscore.com*. 3 November 2010. . Retrieved 24 December 2010.

[134] "Android hits top spot in U.S. smartphone market" (http://news.cnet.com/8301-1035_3-20012627-94.html). 2010-08-04. . Retrieved 2010-08-04.

[135] Greg Sandoval (2010-08-02). "More signs iPhone under Android attack" (http://news.cnet.com/8301-13579_3-20012418-37.html). . Retrieved 2010-08-04.

[136] "Apple iOS Platform Outreaches Android by 59 Percent in U.S. When Accounting for Mobile Phones, Tablets and Other Connected Media Devices" (http://www.comscore.com/Press_Events/Press_Releases/2011/4/). *comScore*. 19 April 2011. .

[137] Dalrymple, Jim (28 April 2011). "The truth about Android vs. iPhone market share" (http://www.loopinsight.com/2011/04/28/the-truth-about-android-vs-iphone-market-share/). .

[138] "Google activates 500,000 Android devices a day, may reach 1 million in October" (http://old.news.yahoo.com/s/digitaltrends/20110628/tc_digitaltrends/googleactivates500000androiddevicesadaymayreach1millionbyoctober). *Yahoo News*. 28 June 2011. .

[139] Barra, Hugo (10 May 2011). "Android: momentum, mobile and more at Google I/O" (http://googleblog.blogspot.com/2011/05/android-momentum-mobile-and-more-at.html). *The Official Google Blog*. . Retrieved 10 May 2011.

[140] Kumparak, Greg (14 July 2011). "Android Now Seeing 550,000 Activations Per Day" (http://techcrunch.com/2011/07/14/android-now-seeing-550000-activations-per-day/). *Techcrunch*. .

[141] "Android takes almost 50% share of worldwide smart phone market" (http://www.canalys.com/newsroom/android-takes-almost-50-share-worldwide-smart-phone-market). 1 August 2011. . Retrieved 2011-08-05.

[142] Erick Schonfeld (13 October 2011). "Larry Page: Mobile Revenues At $2.5 Billion Run-Rate, 190 Million Android Devices" (http://techcrunch.com/2011/10/13/page-google-plus-40-million-mobile-2-5-billion/). *TechCrunch*. . Retrieved 24 October 2011.

[143] Lance Whitney (2011-11-17). "Google: 200 million Android devices now active worldwide" (http://news.cnet.com/8301-1023_3-57326649-93/google-200-million-android-devices-now-active-worldwide/). *CNET News*. . Retrieved 2011-11-27.

[144] Charlie Sorrel (2011-11-17). "Only 3.8 Million Honeycomb Tablets Sold So Far" (http://www.wired.com/gadgetlab/2011/10/only-3-8-million-honeycomb-tablets-sold-so-far/). Wired.com. . Retrieved 2011-11-27.

[145] "Android Platform Versions" (http://developer.android.com/resources/dashboard/platform-versions.html). *Android Developers*. 2012-01-03. . Retrieved 2012-01-03.

[146] "Google Keynote at AnDevCon II" (https://www.youtube.com/watch?v=gIIni_4fC60). *Development Team*. 2011-11-09. .

[147] "Meet Androidland: Australia opens "world-first" Android store in Melbourne" (http://apcmag.com/meet-androidland-australia-opens-world-first-android-store-in-melbourne.htm). Apcmag.com. . Retrieved 2011-12-10.

[148] "Android Open Source Project Frequently Asked Questions: Compatibility" (http://source.android.com/faqs.html#compatibility). *source.android.com*. . Retrieved 13 March 2011.

[149] *Skyhook Wireless, Inc. vs Google, Inc* (15 September 2010) ("This entirely subjective review, conducted solely by Google employees with ultimate authority to interpret the scope and meaning of the CDD as they see fit, effectively gives Google the ability to arbitrarily deem any software, feature or function 'non-compatible' with the CDD."). Text (http://daringfireball.net//misc/2010/09/Skyhook-Google Complaint and Jury Demand.pdf)

[150] "Skyhook Wireless, Inc. vs. Google, Inc." (http://www.socialaw.com/slip.htm?cid=20416&sid=121). *Social Law Library Research Portal*. December 2010. . Retrieved 13 March 2011.

[151] Boulton, Clint (21 October 2008). "Google Open-Sources Android on Eve of G1 Launch" (http://www.eweek.com/c/a/Mobile-and-Wireless/Google-Open-Sources-Android-on-Eve-of-G1-Launch/). *eWeek*. . Retrieved 2009-09-03.

[152] Bort, Dave (21 October 2008). "Android is now available as open source" (http://source.android.com/posts/opensource). *Android Open Source Project*. . Retrieved 2009-09-03.. Mirror link (https://sites.google.com/a/android.com/opensource/posts/opensource).

[153] "Licenses: Android Open Source" (http://source.android.com/source/licenses.html). *Android Open Source Project*. . Retrieved 25 October 2011.

[154] Ryan Paul (2008). "Why Google chose the Apache Software License over GPLv2 for Android" (http://arstechnica.com/old/content/2007/11/why-google-chose-the-apache-software-license-over-gplv2.ars). *Ars Technica*. . Retrieved 25 October 2011.

[155] "Android issues reviewed" (http://code.google.com/p/android/issues/list?q=status:Reviewed). Code.Google.com. . Retrieved 2011-08-08.

[156] "Frequently Asked Questions: What is involved in releasing the source code for a new Android version?" (http://source.android.com/faqs.html#what-is-involved-in-releasing-the-source-code-for-a-new-android-version). *Android Open Source Project*. . Retrieved 25 October 2011.

[157] "Google Android 3.0 "Honeycomb": Open source no more" (http://www.zdnet.com/blog/google/google-android-30-honeycomb-open-source-no-more/2845). ZDNet. 2011-03-24. . Retrieved 2011-07-10.

[158] Bray, Tim (2011-04-06). "Android Developers Blog: I think I'm having a Gene Amdahl moment" (http://android-developers.blogspot.com/2011/04/i-think-im-having-gene-amdahl-moment.html). Android-developers.blogspot.com. . Retrieved 2011-08-08.

[159] Honeycomb won't be open-sourced? Say it ain't so! (2011-03-24). "Honeycomb won't be open-sourced? Say it ain't so!" (http://www.androidcentral.com/google-not-open-sourcing-honeycomb-says-bloomberg). Androidcentral.com. . Retrieved 2011-08-08.

[160] Thom Holwerda (14 November 2011). "Android 4.0 Ice Cream Sandwich Source Code Released" (http://www.osnews.com/story/25330/Android_4_0_Ice_Cream_Sandwich_Source_Code_Released). *OSNews*. . Retrieved 28 November 2011.

[161] Niccolai, James (12 August 2010). "Update: Oracle sues Google over Java use in Android" (http://www.computerworld.com/s/article/9180678/Update_Oracle_sues_Google_over_Java_use_in_Android). *Computerworld*. International Data Group Inc. . Retrieved 20 August 2010.

[162] http://worldwide.espacenet.com/textdoc?DB=EPODOC&IDX=US5966702
[163] http://worldwide.espacenet.com/textdoc?DB=EPODOC&IDX=US6910205
[164] "Oracle's complaint against Google for Java patent infringement" (http://www.scribd.com/doc/35811761/Oracle-s-complaint-against-Google-for-Java-patent-infringement). scribd.com. . Retrieved 2010-08-13.
[165] Singel, Ryan (5 October 2010). "Calling Oracle Hypocritical, Google Denies Patent Infringement" (http://www.wired.com/epicenter/2010/10/google-oracle-android/). *Wired News*. Condé Nast. . Retrieved 26 December 2010.
[166] "Google Answers Oracle, Counterclaims, and Moves to Dismiss Copyright Infringement Claim" (http://groklaw.net/article.php?story=20101005114201136). *Groklaw*. Pamela Jones. 5 October 2010. . Retrieved 26 December 2010.
[167] "Google Files Sizzling Answer to Oracle's Amended Complaint and its Opposition to Motion to Dismiss; updated 2Xs" (http://groklaw.net/article.php?story=20101111114933605). *Groklaw*. 11 November 2010. . Retrieved 26 December 2010.
[168] "Microsoft collects license fees on 50% of Android devices, tells Google to "wake up"" (http://arstechnica.com/microsoft/news/2011/10/microsoft-collects-license-fees-on-50-of-android-devices-tells-google-to-wake-up.ars). *Ars Technica*. . Retrieved 2011-10-24.
[169] Mikael Ricknäs (28 September 2011). "Microsoft signs Android licensing deal with Samsung" (http://www.computerworld.com/s/article/9220357/Microsoft_signs_Android_licensing_deal_with_Samsung). *Computerworld*. . Retrieved 23 October 2011.
[170] "Google publicly accuses Apple, Microsoft, Oracle of patent bullying" (http://arstechnica.com/tech-policy/news/2011/08/google-publicly-accuses-apple-microsoft-oracle-of-patent-bullying.ars). . Retrieved 2011-09-28.
[171] "Google, needing patents, buys Motorola wireless for $12.5 billion" (http://arstechnica.com/gadgets/news/2011/08/google-to-buy-motorola-in-effort-to-defend-itself-from-patent-bullies.ars). . Retrieved 2011-09-28.
[172] Paul, Ryan (January 4, 2012). "Google buys another round of IBM patents as its Oracle trial nears" (http://arstechnica.com/gadgets/news/2012/01/google-buys-another-round-of-ibm-patents-as-oracle-trial-nears.ars). Ars Technica (http://arstechnica.com). . Retrieved January 22, 2012.
[173] Greene, Jay (2012-01-03), *Google's acquisition of IBM patents may aid its Oracle case* (http://news.cnet.com/8301-1023_3-57351633-93/googles-acquisition-of-ibm-patents-may-aid-its-oracle-case/), News.cnet.com, , retrieved 2012-01-10

External links

- Official website (http://http://www.android.com/)
- Android devices at Google.com (http://www.google.com/phone/)
- Google's Android apps (http://www.google.com/mobile/android/)
- Android (operating system) (http://www.dmoz.org/Computers/Systems/Handhelds/Android/) at the Open Directory Project
- Sergey Brin introduces the Android platform (https://www.youtube.com/watch?v=1FJHYqE0RDg) on YouTube
- Android: Building a Mobile Platform to Change the Industry (http://www.stanford.edu/class/ee380/Abstracts/071128.html): lecture given by Google Mobile Platforms Manager, Richard Miner at Stanford University (video archive (http://ee380.stanford.edu/cgi-bin/videologger.php?target=071128-ee380-300.asx))
- Android Internals: Fragment of a course detailing the architecture of Android and interaction of its components (http://technologeeks.com/Courses/Android-Excerpt.pdf)
- Diagram of Android internals (http://www.makelinux.net/android/internals/)

Smartphone

A **smartphone** is a high-end mobile phone built on a mobile computing platform, with more advanced computing ability and connectivity than a contemporary feature phone.[1] [2] [3] The first smartphones were devices that mainly combined the functions of a personal digital assistant (PDA) and a mobile phone or camera phone. Today's models also serve to combine the functions of portable media players, low-end compact digital cameras, pocket video cameras, and GPS navigation units. Modern smartphones typically also include high-resolution touchscreens, web browsers that can access and properly display standard web pages rather than just mobile-optimized sites, and high-speed data access via Wi-Fi and mobile broadband.

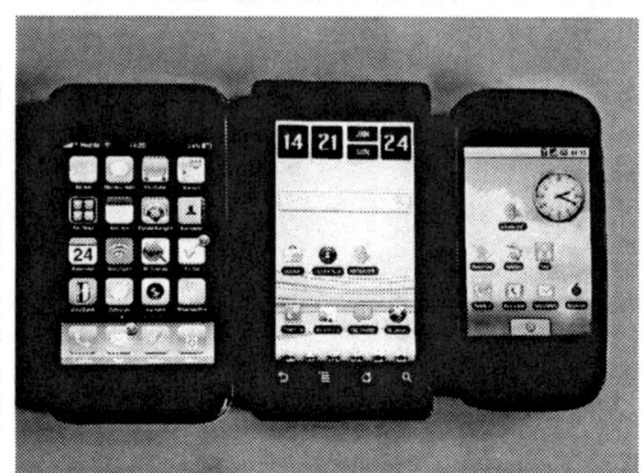

Modern smartphones.

The most common mobile operating systems (OS) used by modern smartphones include Google's Android, Apple's iOS, Microsoft's Windows Phone, Nokia's Symbian, RIM's BlackBerry OS, and embedded Linux distributions such as Maemo and MeeGo. Such operating systems can be installed on many different phone models, and typically each device can receive multiple OS software updates over its lifetime.

The distinction between smartphones and feature phones can be vague and there is no official definition for what constitutes the difference between them. One of the most significant differences is that the advanced application programming interfaces (APIs) on smartphones for running third-party applications[4] can allow those applications to have better integration with the phone's OS and hardware than is typical with feature phones. In comparison, feature phones more commonly run on proprietary firmware, with third-party software support through platforms such as Java ME or BREW.[1] An additional complication in distinguishing between smartphones and feature phones is that over time the capabilities of new models of feature phones can increase to exceed those of phones that had been promoted as smartphones in the past.

History

Early years

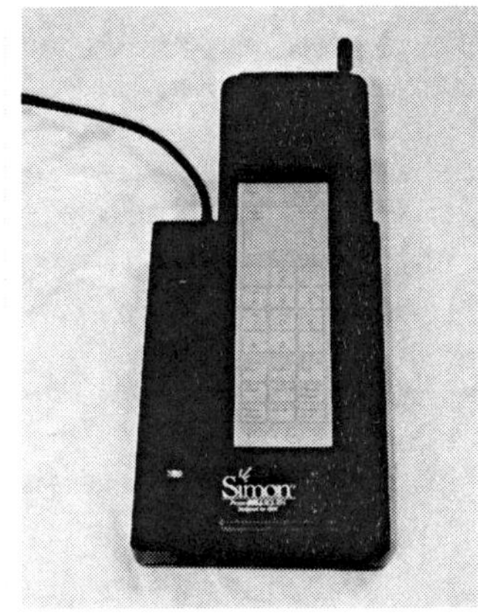

IBM Simon (introduced 1992) shown in the charging station

The first smartphone was the IBM Simon; it was designed in 1992 and shown as a concept product[5] that year at COMDEX, the computer industry trade show held in Las Vegas, Nevada. It was released to the public in 1993 and sold by BellSouth. Besides being a mobile phone, it also contained a calendar, address book, world clock, calculator, note pad, e-mail client, the ability to send and receive faxes, and games. It had no physical buttons, instead customers used a touchscreen to select telephone numbers with a finger or create faxes and memos with an optional stylus. Text was entered with a unique on-screen "predictive" keyboard. By today's standards, the Simon would be a fairly low-end product, lacking a camera and the ability to download third-party applications. However, its feature set at the time was highly advanced.

The Nokia Communicator line was the first of Nokia's smartphones starting with the Nokia 9000, released in 1996. This distinctive palmtop computer style smartphone was the result of a collaborative effort of an early successful and costly personal digital assistant (PDA) by Hewlett-Packard combined with Nokia's best-selling phone around that time, and early prototype models had the two devices fixed via a hinge. The Communicators are characterized by a clamshell design, with a feature phone display, keyboard and user interface on top of the phone, and a physical QWERTY keyboard, high-resolution display of at least 640×200 pixels and PDA user interface under the flip-top. The software was based on the GEOS V3.0 operating system, featuring email communication and text-based web browsing. In 1998, it was followed by Nokia 9110, and in 2000 by Nokia 9110i, with improved web browsing capability.

In 1997 the term 'smartphone' was used for the first time when Ericsson unveiled the concept phone GS88,[6] [7] the first device labelled as 'smartphone'.[8]

Symbian

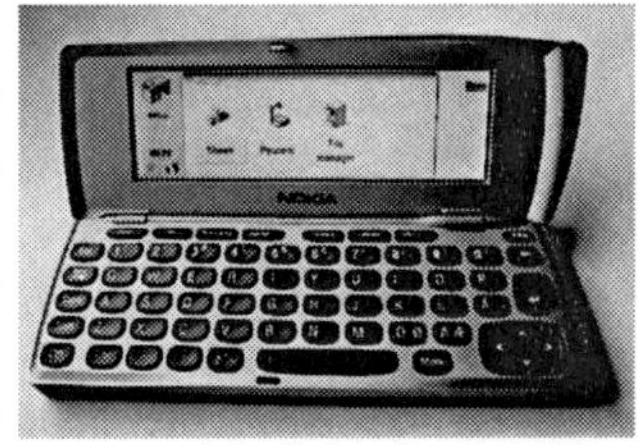

The Nokia 9210 Communicator (Symbian 2000 model smartphone)

In 2000, the touchscreen Ericsson R380 Smartphone was released.[9] It was the first device to use an open operating system, the Symbian OS.[10] It was the first device marketed as a 'smartphone'.[11] It combined the functions of a mobile phone and a personal digital assistant (PDA).[12] In December 1999 the magazine Popular Science appointed the Ericsson R380 Smartphone to one of the most important advances in science and technology.[13] It was a groundbreaking device since it was as small and light as a normal mobile phone.[14] In 2002 it was followed up by P800.[15]

Also in 2000, the Nokia 9210 communicator was introduced, which was the first color screen model from the Nokia Communicator line. It was a true smartphone with an open operating system, the Symbian OS. It was followed by the 9500 Communicator, which also was Nokia's first cameraphone and first Wi-Fi phone. The 9300 Communicator was smaller, and the latest E90 Communicator includes GPS. The Nokia Communicator model is remarkable for also having been the most costly phone model sold by a major brand for almost the full life of the model series, costing easily 20% and sometimes 40% more than the next most expensive smartphone by any major producer.

In 2007 Nokia launched the Nokia N95 which integrated a wide range of multimedia features into a consumer-oriented smartphone: GPS, a 5 megapixel camera with autofocus and LED flash, 3G and Wi-Fi connectivity and TV-out. In the next few years these features would become standard on high-end smartphones. The Nokia 6110 Navigator is a Symbian based dedicated GPS phone introduced in June 2007.

In 2010 Nokia released the Nokia N8 smartphone with a stylus-free capacitive touchscreen, the first device to use the new Symbian^3 OS.[16] It featured a 12 megapixel camera with Xenon flash able to record HD video in 720p, described by Mobile Burn as the best camera in a phone,[17] and satellite navigation that Mobile Choice described as the best on any phone.[18] It also featured a front-facing VGA camera for videoconferencing.

Symbian was the number one smartphone platform by market share from 1996 until 2011 when it dropped to second place behind Google's Android OS. In February 2011, Nokia announced that it would replace Symbian with Windows Phone as the operating system on all of its future smartphones.[19] This transition was completed in October 2011, when Nokia announced its first line of Windows Phone 7.5 smartphones, Lumia 710 and 800.[20]

Palm, Windows, and BlackBerry

In the late 1990s the vast majority of mobile phones had only basic phone features and many people who needed functionality beyond that also carried PDA and/or pager type devices running early versions of operating systems such as Palm OS, BlackBerry OS or Windows CE/Pocket PC.[1] Later versions of these systems started integrating cell phone capabilities with their PDA and messaging features and support of third-party applications. Today, high-end devices running these systems are often branded smartphones.

The HTC Touch Pro2 smartphone (May 2009)

In early 2001, Palm, Inc. introduced the Kyocera 6035, the first smartphone to be deployed in widespread use in the United States. This device combined the features of a personal digital assistant (PDA) with a wireless phone that operated on the Verizon Wireless network. For example, a user could select a name from the PDA contact list, and the device would dial that contact's phone number. The device also supported limited web browsing.[21] The device received a very positive reception from technology publications, but the product line never became widespread outside North America.[22]

In 2001 Microsoft announced its Windows CE Pocket PC OS would be offered as "Microsoft Windows Powered Smartphone 2002."[23] Microsoft originally defined its Windows Smartphone products as lacking a touchscreen and offering a lower screen resolution compared to its sibling Pocket PC devices.

In early 2002 Handspring released the Palm OS Treo smartphone, utilizing a full keyboard that combined wireless web browsing, email, calendar, and contact organizer with mobile third-party applications that could be downloaded or synced with a computer.[24]

In 2002 RIM released their first BlackBerry devices with integrated phone functionality and shifted the positioning of their products from 2-way pagers to email-capable mobile phones. The BlackBerry line evolved into the first smartphone optimized for wireless email use and had achieved a total customer base of about 32 million subscribers by December 2009.[25]

In February 2011 Nokia announced a plan to make Microsoft Windows Phone its main operating system of choice for new Nokia smartphones.[19]

iPhone

In 2007, Apple Inc. introduced its first iPhone. It was initially costly, priced at $499 for the cheaper of two models on top of a two year contract. The first mobile phone to use a multi-touch interface, the iPhone was notable for its use of a large touchscreen for direct finger input as its main means of interaction, instead of having a stylus, keyboard, and/or keypad, which were the typical input methods for other smartphones at the time. The iPhone featured a web browser that *Ars Technica* then described as "far superior" to anything offered by that of its competitors.[26] Initially lacking the capability to install native applications beyond the ones built-in to its OS, at WWDC in June 2007 Apple announced that the iPhone would support third-party "web 2.0 applications" running in its web browser that share the look and feel of the iPhone interface.[27] As a result of the iPhone's initial inability to install third-party native applications, some reviewers did not consider the originally released device to accurately fit the definition of a smartphone "by conventional terms."[28] A process called jailbreaking emerged quickly to provide unofficial third-party native applications. The different functions of the iPhone (including a GPS unit, kitchen timer, radio, map book, calendar, notepad, and many others) allowed consumers to replace all of these items.[29]

The original iPhone (June 2007)

In July 2008, Apple introduced its second generation iPhone with a lower list price starting at $199 and 3G support. Released with it, Apple also created the App Store, adding the capability for any iPhone or iPod Touch to officially execute additional native applications (both free and paid) installed directly over a Wi-Fi or cellular network, without the more typical process at the time of requiring a PC for installation. Applications could additionally be browsed through and downloaded directly via the iTunes software client on Macintosh and Windows PCs, rather than by searching through multiple sites across the Internet. Featuring over 500 applications at launch,[30] Apple's App Store was immediately very popular,[31] quickly growing to become a huge success.[32] [33]

In June 2010, Apple introduced iOS 4, which included APIs to allow third-party applications to multitask,[34] and the iPhone 4, which included a 960×640 pixel display with a pixel density of 326 pixels per inch (ppi), a 5 megapixel camera with LED flash capable of recording HD video in 720p at 30 frames per second, a front-facing VGA camera for videoconferencing, a 1 GHz processor, and other improvements.[35] In early 2011 the iPhone 4 became available through Verizon Wireless, ending AT&T's exclusivity of the handset in the U.S.,[36] [37] [38] and allowing the handset's 3G connection to be used as a wireless Wi-Fi hotspot for the first time, to up to 5 other devices.[39] Software updates subsequently added this capability to other iPhones running iOS 4.[40] [41]

The iPhone 4S was announced on October 4, 2011, improving upon the iPhone 4 with a dual core A5 processor, an 8 megapixel camera capable of recording 1080p video at 30 frames per second, World phone capability allowing it to work on both GSM & CDMA networks, and the Siri automated voice assistant.[42] On October 10, Apple announced that over one million iPhone 4Ss had been pre-ordered within the first 24 hours of it being on sale, beating the 600,000 device record set by the iPhone 4,[43] [44] despite the iPhone 4S failing to impress some critics at the announcement[45] [46] due to their expectations of an "iPhone 5" with rumored drastic changes compared to the iPhone 4 such as a new case design and larger screen.[47] Along with the iPhone 4S Apple also released iOS 5 and iCloud, untethering iOS devices from Macintosh or Windows PCs for device activation, backup, and synchronization,[48] along with additional new and improved features.[49]

There are about 35 percent of Americans that have some sort of smartphone. This shows that the market is spreading fast and there are also more capabilities for smartphones because of this spread.[50]

Smartphones are also mainly valuable based on the operating system. For example, the iPhone runs on the iOS and other devices run different operating systems which makes the functionality of these systems different.[51]

Android

The Android operating system for smartphones was released in 2008. Android is an open-source platform backed by Google, along with major hardware and software developers (such as Intel, HTC, ARM, Motorola and Samsung, to name a few), that form the Open Handset Alliance.[52] The first phone to use Android was the HTC Dream, branded for distribution by T-Mobile as the G1. The software suite included on the phone consists of integration with Google's proprietary applications, such as Maps, Calendar, and Gmail, and a full HTML web browser. Android supports the execution of native applications and a preemptive multitasking capability (in the form of services). Third-party apps are available via the Android Market (released October 2008), including both free and paid apps.

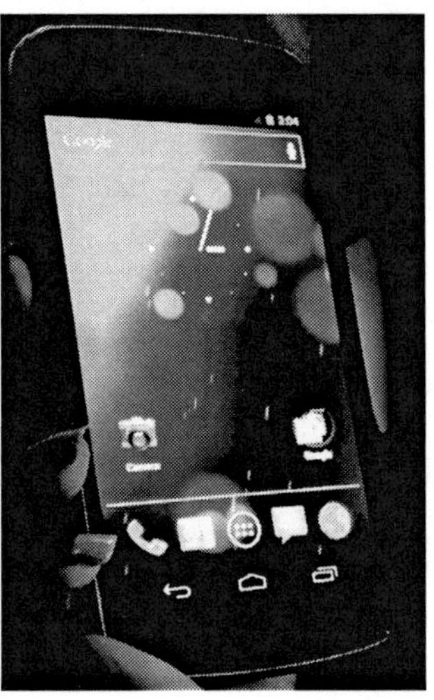
Galaxy Nexus, the latest "Google phone"

In January 2010, Google launched the Nexus One smartphone using its Android OS. Although Android has multi-touch abilities, Google initially removed that feature from the Nexus One,[53] but it was added through a firmware update on February 2, 2010.[54]

Concerning the Xperia Play smartphone, an analyst at CCS Insight said in March 2011 that "Console wars are moving to the mobile platform".[55] In the same month, the HTC EVO 3D was announced by HTC Corporation, which can produce 3D effects with no need for special glasses (autostereoscopy).[56] The HTC EVO 3D was officially released on June 24, 2011.[57]

Bada

The Bada operating system for smartphones was announced by Samsung on 10 November 2009.[58] [59] The first Bada-based phone was the Samsung Wave S8500, released on June 1, 2010,[60] [61] which sold one million handsets in its first 4 weeks on the market.[62]

Samsung shipped 3.5 million phones running Bada in Q1 of 2011.[63] This rose to 4.5 million phones in Q2 of 2011.[64]

Patent licensing and litigation

Recently the number of lawsuits, trade complaints, and countersuits and complaints based on patents and designs in the markets for smartphones, and devices based on smartphone OSes such as Android, has been increasing significantly.

Timeline[65] [66] [67] [68] [69] (initial suits, countersuits, rulings, licence agreements, and other major events in *italics*):

- 2009, Oct 22: *Nokia sues Apple over 10 patents.*[70] [71]
- 2009, Dec 11: *Apple countersues Nokia over 13 patents.*[72]

- 2009, Dec 29: Nokia files a second lawsuit[73] and a U.S. International Trade Commission (ITC) complaint against Apple over 7 more patents.[74]
- 2010, Jan 15: Apple files an ITC complaint against Nokia over 9 patents.[75] [76]
- 2010, Feb 19: Apple drops 4 patents from their countersuit against Nokia that are in their ITC complaint against Nokia.
- 2010, Feb 24: Apple countersues Nokia in Nokia's second lawsuit, over the 9 patents that are in Apple's ITC complaint.
- 2010, Mar 02: *Apple sues HTC over 10 patents and files an ITC complaint against HTC over 10 other patents.*[77] [78] [79] [80] [81]
- 2010, Apr 26: 5 of the patents in Apple's ITC complaint against Nokia are merged into their ITC complaint against HTC.
- 2010, Apr 27: *HTC signs an agreement with Microsoft to licence Microsoft patents in return for royalties on HTC's Android-based devices*[82] [83] (rumored to be $5 per handset).
- 2010, May 7: Nokia files a third lawsuit against Apple over 5 more patents.[84]
- 2010, May 12: *HTC files an ITC complaint against Apple over 5 patents.*[85]
- 2010, May 28: S3 Graphics files an ITC complaint against Apple over 4 patents used in the iPhone, iPod Touch, iPad, and Apple computers.[86]
- 2010, Jun 28: Apple countersues Nokia in Nokia's third lawsuit, over 7 more patents.
- 2010, Jul 06: *HTC countersues Apple over 3 patents.*
- 2010, Jul 21: Nokia drops 1 patent from their ITC complaint against Apple.
- 2010, Aug 12: *Oracle sues Google over 7 patents relating to the use of Java in Android.*[87]
- 2010, Sep 17: Nokia adds 2 more patents to their third lawsuit against Apple.
- 2010, Sep 27: Apple sues Nokia in the UK and Germany over 9 patents.
- 2010, Sep 30: Nokia countersues Apple in Germany over 4 patents.
- 2010, Oct 01: *Microsoft files an ITC complaint and a lawsuit against Motorola over 9 patents.*[88] [89]
- 2010, Oct 06: *Motorola sues Apple over 18 patents, and files an ITC complaint against Apple over 6 of them.*[90]
- 2010, Oct 08: *Motorola files a request for declaratory judgement that they do not infringe 12 Apple patents, and that those patents be declared invalid.*[91] [92]
- 2010, Oct 12: Nokia adds 3 more patents to their countersuit against Apple in Germany.
- 2010, Oct 25: Nokia sues Apple in another German court over 5 patents.
- 2010, Oct 28: Apple drops 4 patents from their ITC complaint against HTC and/or Nokia.
- 2010, Oct 29: *Apple sues Motorola over 6 patents, and files an ITC complaint against Motorola over 3 of them.*[93] [94]
- 2010, Nov 05: HTC drops 1 patent from their ITC complaint against Apple.
- 2010, Nov 09: *Microsoft alleges Motorola has failed to comply with RAND (reasonable and non-discriminatory) licensing obligations.*
- 2010, Nov 10: *Motorola sues Microsoft over 7 patents in one court and 9 patents in another.*
- 2010, Nov 18: *Apple makes counterclaims against Motorola over 6 patents.*
- 2010, Nov 22: *Motorola files an ITC complaint against Microsoft over 5 patents.*
- 2010, Dec 01: *Apple adds the 12 patents to their suit against Motorola that Motorola previously requested declaratory judgement that they do not infringe.*[95]
- 2010, Dec 03: Nokia countersues Apple in the UK over 4 patents, and files a new suit against Apple in the Netherlands over 2 patents.
- 2010, Dec 03: Apple countersues Nokia in Nokia's second German lawsuit, over 1 patent and 2 utility models.
- 2010, Dec 06: Nokia drops 1 patent from their ITC complaint against Apple.
- 2010, Dec 15 and 22: Nokia and Apple take their first German suit/countersuit to the Federal Patent Court of Germany.

- 2010, Dec 23: Motorola files a third lawsuit against Microsoft over 3 patents.
- 2010, Dec 23: Microsoft countersues Motorola over 7 patents.
- 2011, Jan 06: The third Nokia/Apple lawsuit/countersuit is transferred to the location of the first and second ones.
- 2011, Jan 18: Apple seeks to invalidate one Nokia patent in the UK, which it was not yet being sued over.
- 2011, Jan 18: Motorola drops 1 patent from their lawsuits against Microsoft.
- 2011, Jan 19: Microsoft counterclaims against Motorola, asserting 5 patents.
- 2011, Jan 25: Microsoft counterclaims against Motorola, asserting 2 patents.
- 2011, Feb 14: Motorola adds 2 patents to their lawsuits against Microsoft.
- 2011, Feb 22: Apple drops 1 more patent from their ITC complaint against HTC and Nokia.
- 2011, Mar 21: *Microsoft sues Barnes & Noble over the Android operating system in the Nook ebook reader.*[96]
- 2011, Mar 25: *ITC finds that Apple does not infringe on 5 Nokia patents.*
- 2011, Mar 29: Nokia files an ITC complaint against Apple over 7 more patents, and a fourth lawsuit over 6 of those.[97] [98]
- 2011, Apr 15: *Apple sues Samsung for patent and trademark infringement* (7 utility patents, 3 design patents, 3 registered trade dresses, 6 trademarked icons) with its Galaxy line of mobile products, including the *Galaxy S* smartphone and the *Galaxy Tab* tablet.[99] [100]
- 2011, Apr 22: *Samsung sues Apple in South Korea (5 patents), Japan (2 patents), and Germany (3 patents).*[101]
- 2011, Apr 28: *Samsung countersues Apple over 10 patents.*[102]
- 2011, Apr 29: Apple drops 1 more patent from their ITC complaint against HTC.
- 2011, May 18: *Samsung ordered to provide Apple samples of the announced Galaxy S2, Infuse 4G,* and *Infuse 4G LTE* smartphones, as well as the *Galaxy Tab 8.9* and *Galaxy Tab 10.1* tablets as part of Apple's lawsuit against the company.[103] [104]
- 2011, May 18: *Samsung files a court motion for Apple to provide samples of the unannounced iPhone 5 and iPad 3 prototypes.*[105]
- 2011, Jun 14: *Nokia and Apple settle their litigation with Apple agreeing to pay Nokia an undisclosed one-time payment as well as continuing royalties.*[106] [107]
- 2011, Jun 16: *Apple amends its lawsuit against Samsung,* dropping 2 utility patents and 1 design patent, and adding 3 new utility patents plus 4 trade dress applications, *now covering the Samsung Galaxy Tab 10.1*[108]
- 2011, Jun 22: Apple countersues Samsung in South Korea over an unknown number of patents.
- 2011, Jun 22: *Samsung's motion to be provided samples of Apple's unannounced iPhone 5 and iPad 3 prototypes is denied.*[109]
- 2011, Jun 27: General Dynamics Itronix signs an agreement with Microsoft to licence Microsoft patents in return for royalties on General Dynamics Itronix's Android-based devices.[110] [111]
- 2011, Jun 28: Samsung files an ITC complaint and a lawsuit against Apple over 5 patents.
- 2011, Jun 29: Samsung sues Apple in London, UK over an unknown number of patents, and a Samsung lawsuit against Apple in Italy becomes known (details unknown).
- 2011, Jun 29: Velocity Micro signs an agreement with Microsoft to licence Microsoft patents in return for royalties on Velocity Micro's Android-based devices.[111] [112]
- 2011, Jun 30: Samsung converts its countersuit against Apple into counterclaims against Apple's suit, dropping 2 patents but adding 4 more.
- 2011, Jun 30: *A consortium of companies made up of Apple, EMC Corporation, Ericsson, Microsoft, Research In Motion and Sony win against Google*[113] *in an auction of over 6,000 Nortel mobile-related telecommunications patents for $4.5 billion USD.*[114] [115]
- 2011, Jun 30: Onkyo signs an agreement with Microsoft to licence Microsoft patents in return for royalties on Onkyo's Android-based devices.[111] [116]
- 2011, Jul 01: *Apple files for preliminary injunction against 4 Samsung products: Infuse 4G, Galaxy S 4G, Droid Charge,* and *Galaxy Tab 10.1* based on 3 design patents and 1 utility patent.[117]

- 2011, Jul 01: *ITC rules that Apple infringes on 2 patents held by S3 Graphics, while not infringing on 2 others.*[118]
- 2011, Jul 05: *Apple files an ITC complaint against Samsung over 6 smartphones and 2 tablets* infringing 5 utility patents and 2 design patents.
- 2011, Jul 05: Wistron signs an agreement with Microsoft to licence Microsoft patents in return for royalties on Wistron's Android-based devices.[111] [119] [120]
- 2011, Jul 06: *HTC agrees to purchase S3 Graphics* to secure 235 patents for use in its defense against Apple.[121] [122] [123]
- 2011, Jul 06: *Microsoft seeks $15 licensing fees from Samsung for a range of claimed patent violations on every Android device.*[124]
- 2011, Jul 11: Apple files a second ITC complaint against HTC over 5 more patents, and sues HTC over 4 patents from this second ITC complaint that they weren't already suing HTC over.[125] [126]
- 2011, Jul 11-12: Google acquires 1,029 Patents from IBM for an undisclosed amount.[127] [128]
- 2011, Jul 15: *ITC finds HTC infringes on 2 Apple patents.*[129]
- 2011, Jul 29: HTC sues Apple in London, UK over an unknown number of patents.
- 2011, Aug 02: *Apple sues Samsung in Australia over 10 patents, resulting in Samsung delaying the launch and halting advertising of the Samsung Galaxy Tab 10.1 tablet in Australia* to an indefinite date.[130] [131]
- 2011, Aug 09: *A German court issues a preliminary injunction against the Samsung Galaxy Tab 10.1* in Apple's lawsuit against Samsung *which causes its sale to be banned in most of Europe.*[132] [133]
- 2011, Aug 15: *Google announces its intention to buy Motorola Mobility for $12.5 billion USD.* Eighteen of Motorola's patents could potentially be used for defense or countersuits against Apple and Microsoft, and may influence the smartphone war. These patents may change the balance of power, and force the various players to settle their lawsuits.[134] [135]
- 2011, Aug 16: *The Samsung Galaxy Tab 10.1 sales ban in Europe is lifted outside of Germany.*[136] [137]
- 2011, Aug 17: Google acquires 1,023 more patents from IBM for an undisclosed amount (not revealed until 13 Sep 2011).[138]
- 2011, Aug 23: *Microsoft files a complaint with the ITC requesting a ban on several key Motorola smartphones and devices in the USA* based on infringements of 7 patents.[139] [140]
- 2011, Aug 24: *A court in the Netherlands rules that Samsung will be banned from selling the Galaxy S, Galaxy S II and Galaxy Ace in a number of European countries* due to Apple's patent infringement claims.[141]
- 2011, Sep 02: *Apple granted preliminary injunction against Samsung preventing display of the prototype Samsung Galaxy Tab 7.7 tablet at the IFA trade show in Berlin.*[142]
- 2011, Sep 02: *Apple court filings assert that Andy Rubin got inspiration for Android framework while working at Apple* before working at General Magic and Danger, Inc.[143]
- 2011, Sep 07: *HTC countersues Apple using nine patents from Google.* The move is seen as a possible first step for Google giving direct support in lawsuits involving manufacturers using Android.[144] [145] [146] [147]
- 2011, Sep 08: *Acer*[148] *and ViewSonic*[149] *sign patent license agreements with Microsoft* regarding their use of Android on smartphones and tablets.[150] [151]
- 2011, Sep 09: *Apple's preliminary injunction against sales of the Samsung Galaxy Tab 10.1 in Germany is upheld.*[152]
- 2011, Sep 12: Samsung announces a lawsuit against Apple in France that had been filed in July over 3 patents.[153]
- 2011, Sep 12: Apple countersues Samsung in the UK over an unknown number of patents.[154]
- 2011, Sep 13: Google's August 17 acquisition of 1,023 patents from IBM is revealed by the U.S. Patent and Trademark Office.[138] [155]
- 2011, Sep 17: Samsung countersues Apple in Australia over 7 patents.[156]

- 2011, Sep 28: *Samsung signs an agreement with Microsoft to licence Microsoft patents in return for royalties on Samsung's Android-based devices.*[157] [158] [159]
- 2011, Oct 12: *An Australian court issues a preliminary injunction against the Samsung Galaxy Tab 10.1 in Apple's lawsuit against Samsung which prevents its sale in Australia leading up to the 2011 holiday season.*[160]
- 2011, Oct 13: *Quanta signs an agreement with Microsoft to licence Microsoft patents in return for royalties on Quanta's Android and Chrome-based devices.*[161] [162]
- 2011, Oct 13: *Judge in Apple's U.S. lawsuit against Samsung agrees that Samsung's tablets infringe on Apple's patents, but also that the validity of some of the patents might be questionable.*[163]

Screen

Screens on smartphones vary largely in both display size and display resolution. The most common screen sizes range from 2 inches to over 4 inches (measured diagonally). Some 5 inch screen devices exist that run on mobile OSes and have the ability to make phone calls, such as the discontinued Dell Streak and the current Samsung Galaxy Note. Ergonomics arguments have been made that increasing screen sizes start to negatively impact usability.

Common resolutions for smartphone screens vary from 240×320 to 720×1280, with many flagship Android phones at 480×800 or 540×960, the iPhone 4/4S at 640×960 and Galaxy Nexus and HTC Rezound at 720×1280.

Application stores

The introduction of Apple's App Store for the iPhone and iPod Touch in July 2008 popularized manufacturer-hosted online distribution for third-party applications focused on a single platform. Before this, smartphone application distribution was largely dependent on third-party sources providing applications for multiple platforms, such as GetJar, Handango, Handmark, PocketGear, and others.

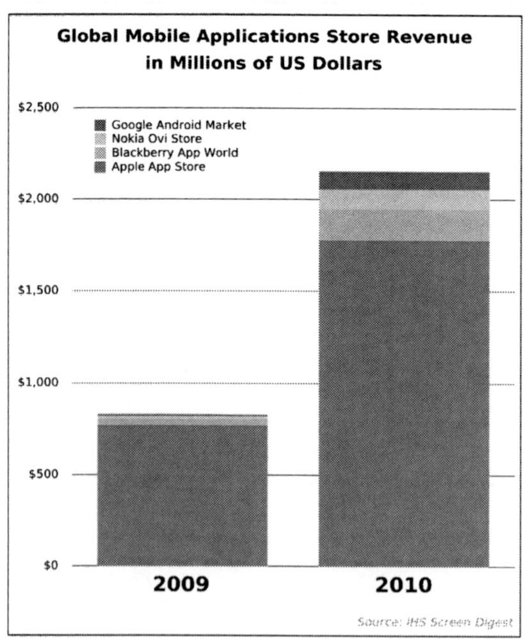

The iPhone's platform is officially restricted to installing apps through the App Store, through "B2B" deployment, and on an "Ad Hoc" basis on up to 100 iPhones.[164] Through jailbreaking it can install apps from other sources. Other platforms may allow application distribution through additional sources outside of their manufacturer-provided app stores, such as third-party app stores and downloads from individual websites.

Following the success of Apple's App Store other smartphone manufacturers quickly launched application stores of their own. Google launched the Android Market in October 2008. RIM launched its app store, BlackBerry App World, in April 2009. Nokia launched its Ovi Store in May 2009. Palm launched its Palm App Catalog for webOS in June 2009. Microsoft launched an application store for Windows Mobile called

Windows Marketplace for Mobile in October 2009, and then a separate Windows Phone Marketplace for Windows Phone in October 2010. Samsung launched Samsung Apps for its Bada based phones in June 2010. Amazon launched its Amazon Appstore for the Google Android operating system in March 2011.

Store	2009 (millions U.S.)	2010 (millions U.S.)[165]
Apple App Store	$769	$1782
Blackberry App World	$36	$165
Nokia Ovi Store	$13	$105
Google Android Market	$11	$102
Total	$828	$2155

The relatively high revenue of U.S. $1782 million in 2010 for Apple's App Store compared to competitor's stores[165] can be attributed to a combination of factors. In large part this can be attributed to having the largest number of apps available and the highest download volume of any mobile app store in 2010, but besides that only 28% of the apps in Apple's App Store were free apps, compared to over 57% in the Android Market. Similarly, Nokia's Ovi Store and the BlackBerry App World both had only 26% of their apps available for free, but both generated higher revenues than the Android Market despite having much lower download volumes.[166]

Malicious software attacks

As smartphone adoption goes up they have increasingly become subject to attacks by malicious software (malware).[167] [168]

Frequently this malware is distributed through application stores that have minimal or no review process for their content.[169] In some cases malware has been hidden in pirated versions of legitimate apps, which are then distributed through 3rd party app stores.[170] [171] Malware risk also comes from what's known as an "update attack," where a legitimate application is later changed to include a malware component, which users then install when they are notified that the app has been updated. Additionally, the ability to acquire software directly from links on the web results in a distribution vector called "malvertizing," where users are directed to click on links, such as on ads that look legitimate, which then open in the device's web browser and cause malware to be downloaded and installed automatically.[172]

Typical smartphone malware leverages platform vulnerabilities that allow it to gain root access on the device in the background. Using this access the malware installs additional software to target communications, location, or other personal identifying information. A common form of malware on mobile phones is the SMS trojan, which sends premium SMS messages, possibly while unknowingly running in the background of a legitimate application. These premium SMS messages run up charges on the owners phone bill which cannot be recovered.

In August 2010, Kaspersky Lab reported detection of the first malicious program for smartphones running on Google's Android operating system, named Trojan-SMS.AndroidOS.FakePlayer.a, an SMS trojan which had already infected a number of devices using that OS.[173] Over the spring of 2011 Android malware increased 76%, according to McAfee.[167] [174] A report from Juniper Global Threat Center notes that malware on the Android platform increased 400% from 2009 to the summer of 2010, and then saw a 472% increase between July and November 2011.[169] The Juniper report indicates that 55% of Android malware acts as spyware, and 44% are SMS trojans.

While there have been and continue to be potential security flaws in iOS,[175] as of at least August 2011 there were no known malware or spyware apps in Apple's App Store, according to security firm Lookout. There are however commercial spyware applications available, outside the App Store, for jailbroken iOS devices.[172] In June 2011 Symantec's 23-page report "A Window Into Mobile Device Security" characterized (non-jailbroken) devices running iOS as having "full protection" against malware attacks.[176]

Symbian and older versions Windows Mobile have had to contend with a degree of malware in the past, but as legacy systems it is believed that the people who previously targeted them have shifted their focus to Android.[169] There were also a few Palm OS viruses.

The only mobile platform other than Apple's iOS without reports of malware so far is HP's (formerly Palm's) webOS, but this may be explained by its relatively low adoption rate.[174]

The best way to reduce a device's vulnerability to malware attacks is to install the most recent versions of operating systems which include security patches. This can be complicated by long delays[177] in software updates for many devices which have had their software modified with custom "skins," services, or promotional on-deck apps by their manufacturer or mobile carrier.[172] In some cases a device may no longer be receiving updates from its manufacturer or carrier, leaving it vulnerable to exploits that have been patched in an OS version that's more recent than the device's last supported one.

Market share

Smartphone market share

For several years, demand for advanced mobile devices boasting powerful processors and graphics processing units, abundant storage (flash memory) for applications and media files, high-resolution screens with multi-touch capability, and open operating systems has outpaced the rest of the mobile phone market.[178]

According to an early 2010 study by ComScore, over 45.5 million people in the United States owned smartphones out of 234 million total subscribers.[179] Despite the large increase in smartphone sales in the last few years, smartphone shipments only made up 20% of total handset shipments as of the first half of 2010.[180]

According to Gartner in their report dated November 2010, total smartphone sales doubled in one year and now smartphones represent 19.3 percent of total mobile phone sales.[181] Smartphone sales increased in 2010 by 72.1 percent from the prior year, whereas sales for all mobile phones only increased by 32%.[182] [183]

According to an Olswang report in early 2011, the rate of smartphone adoption is accelerating: as of March 2011 22% of UK consumers had a smartphone, with this percentage rising to 31% amongst 24- to 35-year-olds.[184]

In March 2011, Berg Insight reported data that showed global smartphone shipments increased 74% from 2009 to 2010.[185]

A survey of mobile users in the United States by Nielsen in Q3, 2011 reports that smartphone ownership has reached 43% of all U.S. mobile subscribers, with the vast majority of users under the age of 44 owning one. In the 25-34 age range smartphone ownership is reported to be at 62%.[186] NPD Group reports that the share of handset sales that were smartphones in Q3, 2011 reached 59% for consumers 18 and over in the U.S.[187]

In profit share worldwide smartphones now far exceed the share of non-smartphones. According to a November 2011 research note from Canaccord Genuity, Apple Inc. holds 52% of the total mobile industry's operating profits, while only holding 4.2% of the global handset market. HTC and RIM similarly only make smartphones and their wordwide profit shares are at 9% and 7%, respectively. Samsung, in second place after Apple at 29%, makes both smartphones and feature phones and doesn't report a breakdown separating their profits between the two kinds of devices, but it can be intuited that a significant portion of that profit comes from their flagship smartphone devices.[188]

Up to the end of November 2011, camera-equipped smartphones took 27 percent of photos, a significant increase from 17 percent last year. Due to the fact that we carry smartphones with us all the time, smartphones have replaced some functions of Point-and-shoot cameras, except the cameras with big optical zoom such as 10x.[189]

Operating system market shares

2010 saw the rapid rise of the Google Android operating system from 4 percent of new deployments in 2009 to 33 percent at the beginning of 2011 making it share the top position with the since long dominating Symbian OS. The smaller rivals include US popular Blackberry OS, iOS, Samsung's recently introduced Bada, HP's heir of Palm webOS and the Microsoft Windows Phone OS which is now supported by Nokia.

Quantity market shares by Gartner (in one year) (new sales)

Smartphone OS	Percent
Android 2009	3.9%
Android 2010	22.7%
Symbian 2009	46.9%
Symbian 2010	37.6%
RIM 2009	19.9%
RIM 2010	16.0%
iOS 2009	14.4%
iOS 2010	15.7%
Windows 2009	8.7%
Windows 2010	4.2%
Other 2009	6.1%
Other 2010	3.8%

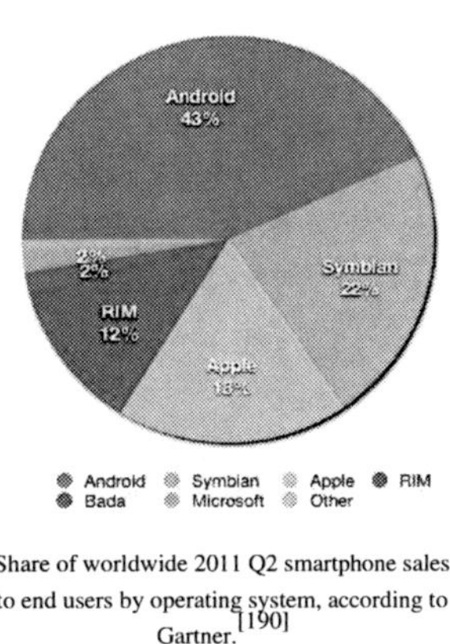

Share of worldwide 2011 Q2 smartphone sales to end users by operating system, according to Gartner.[190]

Over late 2009 and 2010 Android's smartphone operating system market share increased very rapidly.[181] In the fourth quarter of 2010, Android surpassed Symbian as the most common operating system in smartphones, with 32.9 million units sold versus 31.0 million. Android-equipped phones sold seven times more than in the prior year.[191] According to Canalys, Google's Android operating system, which is offered to phone makers for free, has raced to the top past operating systems by Nokia, Apple, RIM, and Microsoft. In Q1 2011 Google's Android market share was 35 percent, increasing significantly from 10 percent the previous year, while Nokia's Symbian dropped to 26 percent from 46 percent over the same time period.[192] In the UK, which currently has one of the highest penetrations of smartphones in the World, Android achieved 50% market share in October 2011.[193]

Historical sales figures

Figures in millions.

Year	Android (Google)	Blackberry (RIM)	IPhone (Apple)	Linux	Palm/WebOS (Palm/HP)	Symbian (Nokia)	Windows Mobile/Phone (Microsoft)
2007[194]		11.77	3.3	11.76	1.76	77.68	14.7
2008[195]		23.15	11.42	11.26	2.51	72.93	16.5
2009[196]	6.8	34.35	24.89	8.13	1.19	80.88	15.03
2010[197]	67.22	47.45	46.6			111.58	12.38

Enterprise share by operating system

In a worldwide study of 2,300 workers at 1,100 businesses by iPass it was reported that Apple's iPhones have displaced RIM's BlackBerry devices in enterprise adoption in 2011.[198] The share for iPhones increased to 45% from 31.1% in 2010, while the Blackberry share dropped to 32.2% from 34.5% in the previous year. Android phones also increased in share, to 21.3% from 11.3% in 2010, exceeding Symbian for the first time, which dropped to 7.4% from 12.4%. Windows Mobile and all other smartphone OSes also dropped in 2011 compared to 2010.[199]

Customer loyalty by operating system

According to a survey of more than 6,000 smartphone users through 2010 by mobile analytics firm Zokem, the top five loyalty scores for smartphone platforms are the iPhone at 73%, followed by Google's Android at 40%, Samsung's Bada at 33%, RIM's BlackBerry at 30%, and Symbian S60 at 23%. Windows Mobile and Palm follow at 10% each. Customer loyalty gauges the likelihood that the user of a smartphone platform whose contract has expired or who has broken or lost their phone will repurchase another one that uses that same platform.[200] [201]

Manufacturer market shares

From the launch of their Communicator model in 1996 until 2011 Nokia was dominant in the smartphone market, though has more recently been joined by other competitors in the market. Based on a report by Strategy Analytics, Samsung overtook Nokia in smartphone shipments with an estimated 27.8 million units shipped in Q3 2011[202] (Samsung does not publicly disclose the numbers of their smartphone shipments and sales).

Quantity market shares by Strategy Analytics (note that Gartner instead shows Nokia ahead on Apple sales based on Symbian sales only[203]
(new sales)

Manufacturer	Percent
Apple Q2 2010	13.5%
Apple Q2 2011	18.5%
Samsung Q2 2010	5.0%
Samsung Q2 2011	17.5%
Nokia Q2 2010	38.1%
Nokia Q2 2011	15.2%
Others Q2 2010	43.4%
Others Q2 2011	48.9%

Market share among smartphone manufacturers does not resemble smartphone OS market share numbers due to the differences between the two major smartphone OS sales models: single manufacturer and licensed. Apple's iPhone, Nokia's Symbian, and RIM's BlackBerry smartphones are currently only available from single manufacturers.

Google's Android OS and Microsoft's mobile OSes are platforms that are licensed and used by a variety of manufacturers. As a result, manufacturers of smartphones using licensed OSes all split the total market share of that OS between them, while the total share for a single-manufacturer OS is held by that manufacturer alone.

Note that Nokia's Symbian OS was previously available from several manufacturers under a licensed model, then later predominantly only by Nokia itself more like a single manufacturer model.

Samsung smartphones use a diverse portfolio of operating systems, including their own Bada operating system along with Android and Windows Mobile.[204]

Apple surpassed Nokia worldwide by revenue and profit for the first time in Q2 2011 (though not in market share), with Apple's profit share of the total worldwide smartphone market increasing to 66.3% while Nokia reported a loss.[205]

Between Q2 2010 and Q2 2011 Nokia's worldwide Symbian smartphone sales dropped significantly from 38.1 percent to 15.2 percent, while Samsung smartphone sales increased significantly worldwide from 5% to 17.5%.[206] As of Q1 2011, Nokia had already announced plans to switch to Windows Phone.

Smartphone Customer Satisfaction by J.D. Power and Associates

Manufacturer	Score
Apple 2010	810
Apple 2011	838
HTC 2010	727
HTC 2011	801
Industry Average 2010	753
Industry Average 2011	788
Samsung 2010	724
Samsung 2011	777
Motorola 2010	N/A
Motorola 2011	775
RIM 2010	741
RIM 2011	762
LG 2010	N/A
LG 2011	760
HP/Palm 2010	712
HP/Palm 2011	733
Nokia 2010	720
Nokia 2011	721

Rankings are based on a possible top score of 1000

Nokia still remains the number one company in the worldwide mobile phone market with sales for Q2 2011 of 88.5 million when including feature phone platforms such as S40, compared with 16.7 million smartphones running Symbian.[207]

According to Nielsen in July 2011, in the United States Apple is the top smartphone manufacturer at 28% of the market, with RIM at 20%. Google Android has 39% of the U.S. market as a whole, but this is split between HTC at 14%, Motorola at 11%, Samsung at 8%, and other remaining manufacturers at 6%. HTC's total share of the U.S.

smartphone market actually ties RIM at 20%, since sales of their smartphones running Microsoft's mobile operating systems account for 6% of the total market. Samsung similarly gains 2% of overall U.S. market share due to their sales of Microsoft OS-based smartphones. In contrast to the worldwide market, Nokia's share of U.S. smartphone sales is very small, at only 2%.[208] [209] Nielsen's Q3, 2011 survey of mobile users maintains Apple as the top U.S. smartphone maker with a continued 28% of the market, with RIM dropping from 20% to 18%.[186] While Google Android increased in total operating system share from 39% to 43% of the U.S. market, it remains fragmented amongst many different manufacturers. Over the same quarter Microsoft managed a modest gain from 6% to 7% total U.S. smartphone OS share.

Checks with U.S. carriers by technology analyst firm Canaccord Genuity in April and August 2011 have found that Apple's iPhone 4 has consistently been the top selling device at AT&T and Verizon. In addition, the second most popular spot at AT&T has been maintained by the iPhone 3GS, which was originally released in 2009 (and has never been sold on Verizon). In August 2011 the most popular smartphones on Sprint and T-Mobile in the U.S. were the HTC EVO 3D 4G and HTC Sensation, respectively. The other second most popular smartphones were the Samsung Charge 4G on Verizon, the Motorola Photon 4G on Sprint, and the HTC myTouch 4G Slide on T-Mobile.[210] [211] NPD Group reported that in Q3, 2011 the overall top 5 smartphones by sales across all carriers in the U.S. were, in order: the iPhone 4, iPhone 3GS, HTC EVO 4G, Motorola Droid 3, and Samsung Intensity II.[187]

Currently the vast majority of smartphones are manufactured in China, Taiwan and Mexico, for companies based in the U.S. (Apple, HP, Motorola), South Korea (LG, Samsung), Canada (RIM), Finland (Nokia), Taiwan (HTC) and the U.K. (Sony Ericsson).

In Q3 2011, Samsung became the world's number one smartphone vendor, selling 24 million smartphones. Nokia's Symbian platform remained in second place with 19.5 million Symbian smartphones, and Apple fell to third place, with 17 million phones. Android had 52.5% of the market, with 60.5 million Android phones being sold. In the US, HTC became the largest smartphone vendor.[212]

Customer satisfaction by manufacturer

According to global marketing information services firm J.D. Power and Associates smartphones from Apple Inc. have been consistently[213] ranking highest in customer satisfaction,[214] [215] with a late 2011 score of 838 out of 1000. Based on the responses to their most recent survey of 6,898 smartphone users, Apple was followed in ranking by HTC (801), Samsung (777), Motorola (775), RIM (762), LG (760), Palm (733), and Nokia (721).[216] [217] [218] [219]

Open-source development

The open-source culture has penetrated the smartphone market in several ways. There have been attempts to open source both hardware and software of smartphones.

In February 2010 Nokia made Symbian open source. Thus, most commercial smartphones were based on open-source operating systems. These include those based on Linux, such as Google's Android, Nokia's Maemo, Hewlett-Packard's webOS, and those based on BSD, such as the Darwin-based Apple iOS. Maemo was later merged with Intel's project Moblin to form MeeGo.[220] [221]

Popular services

Location-based check-in services

According to a ComScore report released on May 12, 2011, nearly one in five smartphone users are tapping into check-in services like Foursquare and Gowalla. A total of 16.7 million mobile-phone subscribers used location-based services on their phones in March 2011.[222]

Second screen

The smartphones have introduced a new way of watching television.[223] The second screen is a consequence of the media multitasking which is exploding.[224]

See also

- Camera phone and videophone
- Comparison of smartphones
- List of digital distribution platforms for mobile devices
- Mobile broadband connectivity
- Mobile Internet device (MID) and personal digital assistant (PDA)
- Mobile operating system
- Second screen

References

[1] "Smartphone" (http://www.phonescoop.com/glossary/term.php?gid=131). *Phone Scoop*. . Retrieved 2011-12-15.

[2] "Feature Phone" (http://www.phonescoop.com/glossary/term.php?gid=310). *Phone Scoop*. . Retrieved 2011-12-15.

[3] Andrew Nusca (20 August 2009). "Smartphone vs. feature phone arms race heats up; which did you buy?" (http://www.zdnet.com/blog/gadgetreviews/smartphone-vs-feature-phone-arms-race-heats-up-which-did-you-buy/6836). ZDNet. . Retrieved 2011-12-15.

[4] "Smartphone definition from PC Magazine Encyclopedia" (http://www.pcmag.com/encyclopedia_term/0,2542,t=Smartphone&i=51537,00.asp). *PC Magazine*. . Retrieved 2011-12-15.

[5] Schneidawind, J: "Big Blue unveiling", *USA Today*, November 23, 1992, p. 2B.

[6] "Ericsson GS88 Preview" (http://pws.prserv.net/Eri_no_moto/GS88_Preview.htm). *Eri-no-moto*. . Retrieved 2011-12-15.

[7] "History" (http://www.stockholmsmartphone.org/history/). Stockholm Smartphone. . Retrieved 2011-12-15.

[8] "Ericsson GS88 box" (http://www.stockholmsmartphone.org/wp-content/uploads/penelope-box.jpg). . Retrieved 2011-12-15.

[9] "PDA Review: Ericsson R380 Smartphone" (http://www.geek.com/hwswrev/pda/ericr380/). *Geek.com*. . Retrieved 2011-12-15.

[10] "Symbian Device – The OS Evolution" (http://www.i-symbian.com/wp-content/uploads/2009/11/Symbian_Evolution.pdf) (PDF). *Independent Symbian Blog*. . Retrieved 2011-12-15.

[11] "Ericsson Introduces The New R380e" (http://www.mobilemag.com/2001/09/25/ericsson-introduces-the-new-r380e). *Mobile Magazine*. . Retrieved 2011-12-15.

[12] "Ericsson R380 World, Review & Rating" (http://www.pcmag.com/article2/0,2817,40827,00.asp). *PCMag.com*. . Retrieved 2011-12-15.

[13] *Popular Science, December 1999* (http://books.google.com/books?id=8qSgh_Q-YOkC). *Google Books*. . Retrieved 2011-12-15.

[14] "Ericsson R380 PDA & Phone" (http://www.cellular.co.za/ericsson_r380.htm). *CellularOnline*. . Retrieved 2011-12-15.

[15] "Sony Ericsson P800" (http://www.allaboutsymbian.com/features/item/Sony_Ericsson_P800.php). *All About Symbian*. . Retrieved 2011-12-15.

[16] "Nokia N8 smartphone official site" (http://www.nokia.co.uk/gb-en/products/phone/n8-00/). . Retrieved 2011-12-15.

[17] "Nokia N8 review - best camera phone ever" (http://www.mobileburn.com/review.jsp?Id=11093). Mobileburn.com. 2010-10-05. . Retrieved 2011-12-15.

[18] "Mobile Choice Award: Best Sat-Nav" (http://www.mobilechoiceuk.com/News/mobile+choice+award:+Best+Sat-Nav/5274). Mobile Choice. 22 October 2010. . Retrieved 2011-12-15.

[19] "Nokia, Microsoft in pact to rival Apple, Google - Technology & Science" (http://www.cbc.ca/news/technology/story/2011/02/11/nokia-microsoft-smart-phone-apple-google.html). Associated Press. CBC.ca. 2011-02-11. . Retrieved 2011-12-15.

[20] Velazco, Chris (26 October 2011). "Nokia Debuts Their First Windows Phones: The Lumia 800 and Lumia 710" (http://techcrunch.com/2011/10/26/nokia-debuts-lumia-710-and-lumia-800/). *TechCruch*. . Retrieved 26 October 2011.

[21] "Kyocera QCP 6035 Smartphone Review" (http://www.palminfocenter.com/view_story.asp?ID=1707). Palminfocenter.com. 2001-03-16. . Retrieved 2011-09-07.

[22] Segan, Sascha (2010-03-23). "Kyocera Launches First Smartphone In Years | News & Opinion" (http://www.pcmag.com/article2/0,2817,2361664,00.asp). PCmag.com. . Retrieved 2011-09-07.

[23] "Better Living through Software: Microsoft Advances for the Home Highlighted at Consumer Electronics Show 2002: Microsoft showcases new and improved products -- including "Freestyle," "Mira" and Ultimate TV at International CES 2002" (http://www.microsoft.com/presspass/features/2002/Jan02/01-08msces.mspx). Microsoft.com. 2002-01-08. . Retrieved 2011-09-07.

[24] Stephen H. Wildstrom (November 30, 2001). "Handspring's Breakthrough Hybrid" (http://www.businessweek.com/bwdaily/dnflash/nov2001/nf20011129_0157.htm). Businessweek.com. . Retrieved 2011-12-15.

[25] Kevin McLaughlin (December 17, 2009). "BlackBerry Users Call For RIM To Rethink Service" (http://www.crn.com/news/client-devices/222002587/blackberry-users-call-for-rim-to-rethink-service.htm). CRN.com. . Retrieved 2011-12-15.

[26] "iPhone in depth: the Ars review" (http://arstechnica.com/apple/reviews/2007/07/iphone-review.ars/6). *ArsTechnica*. Condé Nast. 9 July 2007. p. 6. . Retrieved 3 August 2010.

[27] "iPhone to Support Third-Party Web 2.0 Applications" (http://www.apple.com/pr/library/2007/06/11iphone.html). *Press Release*. Apple Inc.. June 11, 2007. . Retrieved December 15, 2008.

[28] "The iPhone is not a smartphone" (http://www.engadget.com/2007/01/09/the-iphone-is-not-a-smartphone/). Engadget.com. 9 January 2007. . Retrieved 11 July 2010.

[29] "Smartphones Can Replace These Everyday Items" (http://simpleorganizedlife.com/smartphones-can-replace-these-everyday-items/). Simple. Organized. Life.. October 10, 2011. . Retrieved 2011-12-15.

[30] "iPhone 3G on Sale Tomorrow" (http://www.apple.com/pr/library/2008/07/10iphone.html). *Press Release*. Apple Inc.. 2008-07-10. . Retrieved 2009-01-17.

[31] "iPhone App Store Downloads Top 10 Million in First Weekend" (http://www.apple.com/pr/library/2008/07/14iPhone-App-Store-Downloads-Top-10-Million-in-First-Weekend.html). *Press Release*. Apple Inc.. July 14, 2008. . Retrieved 2011-12-15.

[32] "Apple's App Store Downloads Top 1.5 Billion in First Year" (http://www.apple.com/pr/library/2009/07/14Apples-App-Store-Downloads-Top-1-5-Billion-in-First-Year.html). *Press Release*. Apple Inc.. July 14, 2009. . Retrieved 2011-12-15.

[33] "Apple's App Store Downloads Top 15 Billion" (http://www.apple.com/pr/library/2011/07/07Apples-App-Store-Downloads-Top-15-Billion.html). *Press Release*. Apple Inc.. July 7, 2011. . Retrieved 2011-12-15.

[34] "Apple now accepting iOS 4 apps, multitasking ahoy" (http://www.engadget.com/2010/06/11/apple-now-accepting-ios-4-apps-multitasking-ahoy/). Engadget.com. 11 June 2010. . Retrieved 2011-12-15.

[35] "Apple Presents iPhone 4" (http://www.apple.com/pr/library/2010/06/07Apple-Presents-iPhone-4.html). *Press Release*. Apple Inc.. 2010-06-07. . Retrieved 2011-07-05.

[36] "Liveblog: The Verizon iPhone" (http://voices.washingtonpost.com/fasterforward/2011/01/liveblog_the_verizon_iphone.html). *The Washington Post*. .

[37] Memmott, Mark (2011-01-11). "It's Official: Verizon Has The iPhone 4 : The Two-Way" (http://www.npr.org/blogs/thetwo-way/2011/01/11/132833078/its-official-verizon-has-iphone-4). NPR. . Retrieved 2011-09-07.

[38] Raice, Shayndi (January 12, 2011). "Verizon Unwraps iPhone" (http://online.wsj.com/article/SB10001424052748703791904576075681886276172.html?mod=googlenews_wsj). *The Wall Street Journal*. .

[39] Glenn Fleishman (2011-02-22). "Using the Personal Hotspot on your Verizon iPhone" (http://www.macworld.com/article/158058/2011/02/personal_hotspot_verizon.html). Macworld.com. . Retrieved 2011-03-12.

[40] "iOS 4.3 Software Update" (http://www.apple.com/ios/). Apple Inc.. . Retrieved 2011-03-12.

[41] Dan Moren (2011-03-11). "Hands on with iOS 4.3" (http://www.macworld.com/article/158483/2011/03/firstlook_43.html). Macworld.com. . Retrieved 2011-03-12.

[42] "Apple Launches iPhone 4S, iOS 5 & iCloud" (http://www.apple.com/pr/library/2011/10/04Apple-Launches-iPhone-4S-iOS-5-iCloud.html). Apple Inc.. October 4, 2011. . Retrieved 2011-12-15.

[43] "iPhone 4S Pre-Orders Top One Million in First 24 Hours" (http://www.apple.com/pr/library/2011/10/10iPhone-4S-Pre-Orders-Top-One-Million-in-First-24-Hours.html). Apple. . Retrieved 10 October 2011.

[44] Anderson, Ash. "iPhone 4S Sells 1 Million in Under 24 Hours" (http://www.keynoodle.com/iphone-4s-sells-1-million-in-under-24-hours/). *KeyNoodle*. . Retrieved 2011-12-15.

[45] "Apple's fall from grace" (http://www.apple.com/pr/library/2011/10/04Apple-Launches-iPhone-4S-iOS-5-iCloud.html). BGR.com. October 5, 2011. . Retrieved 2011-12-15.

[46] Michael E. Cohen (October 13, 2011). "iPhone 4S: A Very Palpable Hit" (http://tidbits.com/article/12554). tidbits.com. . Retrieved 2011-12-15.

[47] David Pogue (October 11, 2011). "New iPhone Conceals Sheer Magic" (http://www.nytimes.com/2011/10/12/technology/personaltech/iphone-4s-conceals-sheer-magic-pogue.html?_r=2&pagewanted=all). The New York Times. . Retrieved 2011-12-15.

[48] "Press Info - Apple to Launch iCloud on October 12" (http://www.apple.com/pr/library/2011/10/04Apple-to-Launch-iCloud-on-October-12.html). Apple. 2011-10-04. . Retrieved 2012-01-05.

[49] "Press Info - New Version of iOS Includes Notification Center, iMessage, Newsstand, Twitter Integration Among 200 New Features" (http://www.apple.com/pr/library/2011/06/06New-Version-of-iOS-Includes-Notification-Center-iMessage-Newsstand-Twitter-Integration-Among-200-New-Features.html). Apple.

2011-06-06. . Retrieved 2012-01-05.

[50] Michael B. Farrell (November 12, 2011). "No cash, card? No problem: Paying by smartphone is an emerging trend" (http://articles.boston.com/2011-11-12/business/30391540_1_smartphone-mobile-commerce-card-readers). boston.com. . Retrieved 2011-12-15.

[51] 13 November 2011. (http://cellphones.about.com/od/smartphonebasics/a/what_is+_smart.htm)

[52] "Alliance Members" (http://www.openhandsetalliance.com/oha_members.html). *Open Handset Alliance*. . Retrieved 16 January 2011.

[53] "Weighing Nexus over iPhone – a practical review for the everyday user!" (http://techietrick.blogspot.com/2010/01/weighing-nexus-over-iphone-practical.html). Techietrick.blogspot.com. 2010-01-06. . Retrieved 2011-09-07.

[54] "Nexus One gets a software update, enables multitouch (updated with video!)" (http://www.engadget.com/2010/02/02/nexus-one-gets-a-software-update-enables-multitouch). Engadget.com. . Retrieved 2011-09-07.

[55] Tarmo Virki (2011-02-14). "Sony takes gaming console war to phones" (http://in.reuters.com/article/2011/02/13/idINIndia-54865020110213). Reuters. .

[56] "HTC EVO 3D" (http://web.archive.org/web/20110511105547/http://www.htc.com/www/product/evo3d/overview.html). htc.com. Archived from the original (http://www.htc.com/www/product/evo3d/overview.html) on 2011-05-11. . Retrieved 2011-12-15.

[57] Savov, Vlad. "HTC Evo 3D Launches June 24th" (http://www.engadget.com/2011/06/06/htc-evo-3d-launches-on-june-24th-for-200-joined-by-evo-view-4g/). Engadget.com. . Retrieved 7 June 2011.

[58] Ed Hansberry (11 November 2009). "Samsung Bailing on Windows Mobile" (http://www.informationweek.com/blog/main/archives/2009/11/samsung_bailing.html). *InformationWeek*. .

[59] "Samsung to Discard Windows Phone" (http://www.telecomskorea.com/market-8281.html). *Telecoms Korea*. 9 November 2009. .

[60] "Samsung Wave, first Bada smartphone hits the market" (http://www.bada.com/samsung-wave-first-bada-smartphone-hits-the-market/). *Bada*. 24 May 2010. . Retrieved 3 February 2011.

[61] "BadaWave" (http://badawave.com/). BadaWave. . Retrieved 2012-01-05.

[62] "Samsung Waves away a million" (http://www.theinquirer.net/inquirer/news/1722287/samsung-waves-away-million). The Inquirer. 13 July 2010. .

[63] "Android increases smart phone market leadership with 35% share" (http://www.canalys.com/newsroom/android-increases-smart-phone-market-leadership-35-share). Canalys.com. 2011-05-04. . Retrieved 2011-09-07.

[64] "Samsung Bada shipments up 355% to 4.5 million units in Q2 2011 | asymco news | PG.Biz" (http://www.pocketgamer.co.uk/r/PG.Biz/asymco+news/news.asp?c=32049). Pocket Gamer. . Retrieved 2011-09-07.

[65] Florian Mueller. "Apple vs Android 10.12.02" (http://www.scribd.com/doc/44759893/Apple-vs-Android-10-12-02). . Retrieved 2011-09-08.

[66] Florian Mueller. "NokiaVsApple_11.03.31.100" (http://www.scribd.com/doc/52195210/NokiaVsApple-11-03-31-100). . Retrieved 2011-09-08.

[67] Florian Mueller. "Microsoft vs Motorola 11.04.09" (http://www.scribd.com/doc/52754079/Microsoft-vs-Motorola-11-04-09). . Retrieved 2011-09-08.

[68] Florian Mueller. "AppleVsSamsung_11.07.05" (http://www.scribd.com/doc/59474521/AppleVsSamsung-11-07-05). . Retrieved 2011-09-08.

[69] Florian Mueller. "AppleVsHTCandS3_11.07.29" (http://www.scribd.com/doc/61387056/AppleVsHTCandS3-11-07-29). . Retrieved 2011-09-08.

[70] "Nokia sues Apple over iPhone's use of patented wireless standards" (http://www.appleinsider.com/articles/09/10/22/nokia_sues_apple_over_iphones_use_of_patented_wireless_standards.html). Appleinsider.com. 2009-10-22. . Retrieved 2012-01-05.

[71] "Nokia vs. Apple: the in-depth analysis" (http://www.engadget.com/2009/10/29/nokia-vs-apple-the-in-depth-analysis/). Engadget.com. . Retrieved 2012-01-05.

[72] "Apple countersues Nokia for infringing 13 patents" (http://www.engadget.com/2009/12/11/apple-countersues-nokia-for-infringing-13-patents/). Engadget.com. . Retrieved 2012-01-05.

[73] Foresman, Chris (2010-01-04). "Nokia adds additional lawsuit in patent catfight with Apple" (http://arstechnica.com/apple/news/2010/01/nokia-adds-additional-lawsuit-in-patent-catfight-with-apple.ars). Arstechnica.com. . Retrieved 2012-01-05.

[74] Foresman, Chris (2009-12-29). "Nokia hurls new salvo in spat with Apple, complains to ITC" (http://arstechnica.com/apple/news/2009/12/nokia-hurls-new-salvo-in-spat-with-apple-complains-to-itc.ars). Arstechnica.com. . Retrieved 2012-01-05.

[75] Decker, Susan (2010-01-16). "Apple Files New Trade Complaint With Against Nokia" (http://www.bloomberg.com/apps/news?pid=newsarchive&sid=ao_5HVbD_IRM). Bloomberg.com. . Retrieved 2012-01-05.

[76] Cheng, Jacqui (2010-01-18). "Apple wants Nokia's US imports blocked" (http://arstechnica.com/apple/news/2010/01/apple-continues-nokia-patent-spat-with-its-own-itc-complaint.ars). Arstechnica.com. . Retrieved 2012-01-05.

[77] "Apple vs HTC: a patent breakdown" (http://www.engadget.com/2010/03/02/apple-vs-htc-a-patent-breakdown/). Engadget.com. 2010-03-02. . Retrieved 2012-01-05.

[78] "Google backs HTC in what could be 'long and bloody battle' with Apple" (http://www.appleinsider.com/articles/10/03/03/google_backs_htc_in_what_could_be_long_and_bloody_battle_with_apple.html). Appleinsider.com. 2010-03-03. . Retrieved 2012-01-05.

[79] Bilton, Nick (2010-03-02). "What Apple vs. HTC Could Mean" (http://bits.blogs.nytimes.com/2010/03/02/what-apple-vs-htc-could-mean/). Bits.blogs.nytimes.com. . Retrieved 2012-01-05.

[80] "HTC says it uses own technology, not Apple's" (http://www.macworld.com/article/146819/2010/03/htc_defense.html). Macworld.com. . Retrieved 2012-01-05.

[81] Foresman, Chris (2010-03-09). "HTC lawsuit came after warning by Apple to handset makers" (http://arstechnica.com/apple/news/2010/03/htc-lawsuit-came-after-warning-by-apple-to-handset-makers.ars). Arstechnica.com. . Retrieved 2012-01-05.
[82] "Microsoft Announces Patent Agreement With HTC" (http://www.microsoft.com/Presspass/press/2010/apr10/04-27MSHTCPR.mspx). Microsoft.com. 2010-04-27. . Retrieved 2012-01-05.
[83] "HTC licenses Microsoft's mobile patents for Android phones – First Take" (http://gartenberg.wordpress.com/2010/04/28/htc-licenses-microsofts-mobile-patents-for-android-phones-first-take/). Gartenberg.wordpress.com. 2010-04-28. . Retrieved 2012-01-05.
[84] Johnston, Casey (2010-05-07). "Nokia applies patent thumbscrews to Apple's iPad" (http://arstechnica.com/apple/news/2010/05/nokia-applies-patent-thumbscrews-to-apples-ipad.ars). Arstechnica.com. . Retrieved 2012-01-05.
[85] "HTC countersues Apple, claims infringement of five patents" (http://www.appleinsider.com/articles/10/05/12/htc_countersues_apple_claims_infringement_of_five_patents.html). Appleinsider.com. 2010-05-12. . Retrieved 2012-01-05.
[86] "S3 Graphics Files New 337 Complaint Regarding Certain Electronic Devices With Image Processing Systems" (http://www.itcblog.com/20100602/s3-graphics-files-new-337-complaint-regarding-certain-electronic-devices-with-image-processing-systems/). Itcblog.com. 2010-06-02. . Retrieved 2012-01-05.
[87] August 12, 2010 (2010-08-12). "Oracle sues Google over Android" (http://venturebeat.com/2010/08/12/oracle-sues-google-over-android/). Venturebeat.com. . Retrieved 2012-01-05.
[88] Protalinski, Emil (2010-10-02). "Microsoft sues Motorola, citing Android patent infringement" (http://arstechnica.com/microsoft/news/2010/10/microsoft-sues-motorola-citing-android-patent-infringement.ars). Arstechnica.com. . Retrieved 2012-01-05.
[89] "Microsoft Files Patent Infringement Action Against Motorola" (http://www.microsoft.com/presspass/press/2010/oct10/10-01statement.mspx). Microsoft.com. 2010-10-01. . Retrieved 2012-01-05.
[90] Foresman, Chris (2010-10-06). "Motorola asks ITC, two federal courts to throw book at Apple" (http://arstechnica.com/apple/news/2010/10/motorola-asks-itc-two-federal-courts-to-throw-book-at-apple.ars). Arstechnica.com. . Retrieved 2012-01-05.
[91] "Motorola Moves Offensively with a New Lawsuit against Apple" (http://www.patentlyapple.com/patently-apple/2010/10/motorola-moves-offensively-with-a-new-lawsuit-against-apple.html). Patentlyapple.com. 2010-10-15. . Retrieved 2012-01-05.
[92] Motorola seeks to invalidate Apple phone patents (http://arstechnica.com/apple/news/2010/10/motorola-horns-in-on-apple-vs-htc-files-suit-against-apple.ars).
[93] Apple Files Lawsuit against Motorola to Defend Multi-Touch (http://www.patentlyapple.com/patently-apple/2010/10/apple-files-lawsuit-against-motorola-to-defend-multi-touch.html).
[94] Apple Patent Case Against Motorola to Get Trade Agency Review (http://www.businessweek.com/news/2010-11-23/apple-patent-case-against-motorola-to-get-trade-agency-review.html).
[95] Apple vs. Motorola: now 42 patents-in-suit (24 Apple and 18 Motorola patents) (http://fosspatents.blogspot.com/2010/12/apple-vs-motorola-now-42-patents-in.html).
[96] Microsoft sues Barnes & Noble over Android in Nook (http://www.geekwire.com/2011/breaking-microsoft-sues-barnes-noble-android).
[97] Nokia files second ITC complaint against Apple (http://press.nokia.com/2011/03/29/nokia-files-second-itc-complaint-against-apple/).
[98] FOSS Patents: Escalation: Nokia files new ITC complaint against Apple (plus a corresponding federal lawsuit) (http://fosspatents.blogspot.com/2011/03/escalation-nokia-files-new-itc.html).
[99] "Apple to Samsung: Stop stealing ideas" (http://www.cnn.com/2011/TECH/innovation/04/19/apple.samsung.lawsuit.wired/). *CNN.com*. . Retrieved 19 April 2011.
[100] Apple sues Samsung: a complete lawsuit analysis (http://thisismynext.com/2011/04/19/apple-sues-samsung-analysis/).
[101] Samsung Countersues Apple for Patent Infringement (http://www.pcmag.com/article2/0,2817,2383964,00.asp).
[102] Samsung sues Apple for infringing 10 patents: a closer look (http://thisismynext.com/2011/04/29/samsung-sues-apple-infringing-10-patents-closer/).
[103] Samsung Must Show Cell Phones to Apple (http://www.courthousenews.com/2011/05/19/36708.htm).
[104] Judge Orders Samsung to Give Apple Unreleased Tablets, Smartphones (http://www.pcmag.com/article2/0,2817,2385861,00.asp).
[105] Samsung's lawyers demand to see the iPhone 5 and iPad 3 (http://thisismynext.com/2011/05/28/samsung-apple-iphone-5-ipad-3/).
[106] Nokia Wins Apple Patent-License Deal Cash, Settles Lawsuits (http://www.bloomberg.com/news/2011-06-14/nokia-apple-payments-to-nokia-settle-all-litigation.html).
[107] Apple Settles With Nokia In Patent Lawsuit (http://www.huffingtonpost.com/2011/06/14/apple-nokia-patent-lawsuit-settlement_n_876499.html).
[108] Staff Reporter, "Apple alleges Samsung of 'slavishly copying' its technology; questions new Galaxy Tab 10.1" (http://www.ibtimes.com/articles/165405/20110619/apple-lawsuit-upgrade-samsung-galaxy-tab-10-1-infringement-intellectual-property-rights-ipad-2.htm), *International Business Times*, 19 June 2011.
[109] Shocker: Samsung not allowed to see iPhone 5 and iPad 3 (http://thisismynext.com/2011/06/22/shocker-samsung-allowed-iphone-5-ipad-3/).
[110] "Microsoft and General Dynamics Itronix Sign Patent Agreement: Agreement will cover General Dynamics Itronix devices running the Android platform" (http://www.microsoft.com/Presspass/press/2011/jun11/06-27ItronixPR.mspx). Microsoft.com. 2011-06-27. . Retrieved 2012-01-05.
[111] FOSS Patents: Android device makers sign up as Microsoft patent licensees -- an overview of the situation (http://fosspatents.blogspot.com/2011/06/android-device-makers-sign-up-as.html).

[112] "Microsoft and Velocity Micro, Inc., Sign Patent Agreement Covering Android-Based Devices: Agreement provides broad coverage of Microsoft's patent portfolio" (http://www.microsoft.com/Presspass/press/2011/jun11/06-29VelocityMicroPR.mspx). Microsoft.com. 2011-06-29. . Retrieved 2012-01-05.
[113] Dealtalk: Google bid "pi" for Nortel patents and lost (http://www.reuters.com/article/2011/07/02/us-dealtalk-nortel-google-idUSTRE76104L20110702).
[114] Nortel Announces the Winning Bidder of Its Patent Portfolio for a Purchase Price of US$4.5 Billion (http://www.marketwatch.com/story/nortel-announces-the-winning-bidder-of-its-patent-portfolio-for-a-purchase-price-of-us45-billion-2011-06-30?reflink=MW_news_stmp).
[115] Who Won The 6,000+ Nortel Patents? Apple, RIM, Microsoft — Everyone But Google (http://techcrunch.com/2011/07/01/apple-microsoft-rim-google-nortel-patents/).
[116] "Microsoft and Onkyo Corp. Sign Patent Agreement Covering Android-Based Tablets: Agreement provides broad coverage of Microsoft's patent portfolio" (http://www.microsoft.com/Presspass/press/2011/jun11/06-30OnkyoPR.mspx?rss_fdn=Custom). Microsoft.com. . Retrieved 2012-01-05.
[117] Apple files motion for preliminary injunction in the U.S. against four Samsung products: Infuse 4G, Galaxy S 4G, Droid Charge, Galaxy Tab 10.1 (http://fosspatents.blogspot.com/2011/07/apple-files-motion-for-preliminary.html).
[118] ITC ruling mixed in S3 Graphics v. Apple (http://news.cnet.com/8301-27076_3-20076219-248/itc-ruling-mixed-in-s3-graphics-v-apple/?tag=mncol;txt).
[119] "Microsoft and Wistron Sign Patent Agreement: Agreement will cover Wistron's Android tablets, smartphones and e-readers" (http://www.microsoft.com/Presspass/press/2011/jul11/07-05WistronPR.mspx). Microsoft.com. 2011-07-05. . Retrieved 2012-01-05.
[120] Microsoft patent division taking cash from at least 5 Android vendors (http://www.networkworld.com/news/2011/070511-microsoft-patent-android.html).
[121] DailyTech - VIA, WTI Sell Stakes in S3 Graphics to HTC (http://www.dailytech.com/VIA+WTI+Sell+Stakes+in+S3+Graphics+to+HTC/article22078.htm).
[122] HTC to Acquire Chip Designer S3 Graphics, Securing Patents in Apple Fight - Bloomberg (http://www.bloomberg.com/news/2011-07-06/htc-to-buy-s3-graphics-securing-patents-in-apple-fight-1-.html).
[123] HTC facing backlash for S3 acquisition (http://news.cnet.com/8301-1035_3-20077720-94/htc-facing-backlash-for-s3-acquisition/?tag=mncol;txt).
[124] Microsoft wants Samsung to pay smartphone license (http://www.reuters.com/article/2011/07/06/us-samsung-microsoft-idUSTRE7651DB20110706).
[125] HTC's Flyer Tablet, Droid Accused of Infringing Apple Patents - Bloomberg (http://www.bloomberg.com/news/2011-07-11/apple-files-new-trade-complaint-against-htc-over-devices.html?cmpid=yhoo).
[126] FOSS Patents: Apple files second ITC complaint against HTC: better luck next time? (http://fosspatents.blogspot.com/2011/07/apple-files-second-itc-complaint.html).
[127] Google Acquires Over 1,000 IBM Patents in July (http://www.seobythesea.com/2011/07/google-acquires-ibm-patents-in-july/).
[128] Google's New Patents from IBM (http://www.seobythesea.com/2011/07/googles-new-patents-from-ibm/).
[129] Bad News for Android: ITC Rules HTC Violated Two Apple Patents - John Paczkowski - News - AllThingsD (http://allthingsd.com/20110715/itc-rules-htc-violated-two-apple-patents/?refcat=news).
[130] Apple Lawsuit Puts Samsung Tablet Sales in Australia on Hold (http://www.bloomberg.com/news/2011-08-01/apple-seeks-to-block-samsung-from-selling-tablet-in-australia.html).
[131] Samsung's official comment on the Australian Galaxy Tab 10.1 situation is extremely weak (http://fosspatents.blogspot.com/2011/08/samsungs-official-comment-on-australian.html).
[132] Tsukayama, Hayley (10 August 2011). "Samsung Galaxy Tab 10.1 imports banned in most of Europe" (http://www.washingtonpost.com/blogs/faster-forward/post/samsung-galaxy-tab-101-imports-banned-in-most-of-europe/2011/08/10/gIQAmnVr6I_blog.html). *The Washington Post*. . Retrieved 11 August 2011.
[133] Preliminary injunction granted by German court: Apple blocks Samsung Galaxy Tab 10.1 in the entire European Union except for the Netherlands (http://fosspatents.blogspot.com/2011/08/preliminary-injunction-granted-by.html).
[134] "Microsoft files to ban imports of Motorola smartphones" (http://androidandme.com/2011/08/news/microsoft-files-to-ban-imports-of-motorola-smartphones). Android and Me. . Retrieved 2011-12-10.
[135] Womack, Brian (2011-08-22). "Motorola Value Found in 18 Patents Used Against Apple: Tech" (http://www.bloomberg.com/news/2011-08-22/motorola-s-value-for-google-found-in-18-patents-used-against-apple-tech.html). Bloomberg. . Retrieved 2011-12-10.
[136] "Samsung Galaxy tablet ban lifted in most of Europe" (http://uk.reuters.com/article/2011/08/16/tech-us-samsung-apple-ban-idUKTRE77F45420110816). *Reuters*. . Retrieved 16 August 2011.
[137] "Samsung Galaxy Tab ban is on hold" (http://www.bbc.co.uk/news/business-14548895). *BBC*. . Retrieved 16 August 2011.
[138] Google and IBM do it again: Google Acquires over 1,000 Patents from IBM in August (http://www.seobythesea.com/2011/09/google-ibm-patents-august/#more-6675).
[139] Microsoft Asks For An Import Ban On Motorola Smartphones (http://techcrunch.com/2011/08/23/microsoft-asks-for-an-import-ban-on-motorola-smartphones/).
[140] "Microsoft files to ban imports of Motorola smartphones" (http://androidandme.com/2011/08/news/microsoft-files-to-ban-imports-of-motorola-smartphones/). Android and Me. . Retrieved 2011-12-10.

[141] Euro ban for Samsung Galaxy phone (http://www.bbc.co.uk/news/technology-14652482).
[142] Samsung Puts Galaxy 10.1 Tablet on Hold as Apple Wins German Court Order (http://www.bloomberg.com/news/2011-09-04/apple-wins-german-injunction-on-new-galaxy-tab.html).
[143] FOSS Patents: Apple to ITC: Andy Rubin got inspiration for Android framework while working at Apple, hence infringes an Apple API patent (http://fosspatents.blogspot.com/2011/09/apple-to-itc-andy-rubin-got-inspiration.html).
[144] Milford, Phil (2011-09-08). "HTC Sues Apple Using Google Patents Bought Last Week as Battle Escalates" (http://www.bloomberg.com/news/2011-09-07/htc-sues-apple-alleging-infringement-of-four-u-s-patents.html). Bloomberg. . Retrieved 2011-12-10.
[145] Google Hands HTC Patents to Use Against Apple in Smartphone Wars (http://www.businessweek.com/news/2011-09-08/google-hands-htc-patents-to-use-against-apple-in-smartphone-wars.html).
[146] Patel, Nilay (2011-09-07). "HTC sues Apple for patent infringement… using patents purchased by Google" (http://thisismynext.com/2011/09/07/htc-sues-apple-patent-infringement-patents-purchased-google/). Thisismynext.com. . Retrieved 2012-01-05.
[147] Google gets its hands dirty - Apple 2.0 - Fortune Tech (http://tech.fortune.cnn.com/2011/09/08/google-gets-its-hands-dirty/).
[148] "Microsoft and Acer Sign Patent License Agreement: Agreement will cover Acer's Android tablets and smartphones" (http://www.microsoft.com/Presspass/press/2011/sep11/09-08AcerPR.mspx). Microsoft.com. 2011-09-08. . Retrieved 2012-01-05.
[149] "Microsoft and ViewSonic Sign Patent Agreement: Agreement will cover ViewSonic's Android Tablets and smartphones" (http://www.microsoft.com/Presspass/press/2011/sep11/09-08ViewSonicPR.mspx). Microsoft.com. 2011-09-08. . Retrieved 2012-01-05.
[150] "Microsoft ropes two more OEMs into Android patent deal" (http://www.zdnet.com/blog/hardware/microsoft-ropes-two-more-oems-into-android-patent-deal/14622). Zdnet.com. . Retrieved 2012-01-05.
[151] "Microsoft signs Android patent agreements with Acer, ViewSonic" (http://seattletimes.nwsource.com/html/microsoftpri0/2016144500_microsoft_signs_android_patent_agreements_with_ace.html). Seattletimes.nwsource.com. 2011-09-08. . Retrieved 2012-01-05.
[152] Matussek, Karin (2011-09-09). "Apple Wins German Ban on Samsung Tablet" (http://www.bloomberg.com/news/2011-09-09/apple-wins-ruling-for-german-samsung-galaxy-tablet-10-1-ban.html). Bloomberg.com. . Retrieved 2012-01-05.
[153] "Latest Samsung lawsuit targets Apple's iPhone, iPad in France" (http://www.appleinsider.com/articles/11/09/13/latest_samsung_lawsuit_targets_apples_iphone_ipad_in_france.html). AppleInsider. 2011-09-13. . Retrieved 2012-01-05.
[154] Meyer, David (2011-09-14). "Apple sues Samsung in the UK over Android | ZDNet UK" (http://www.zdnet.co.uk/blogs/communication-breakdown-10000030/apple-sues-samsung-in-the-uk-over-android-10024342/). Zdnet.co.uk. . Retrieved 2012-01-05.
[155] "USPTO Assignments on the Web" (http://assignments.uspto.gov/assignments/q?db=pat&reel=026894&frame=0001). Assignments.uspto.gov. . Retrieved 2012-01-05.
[156] "Samsung fires back at Apple in Australia with countersuit against iPhone, iPad" (http://www.appleinsider.com/articles/11/09/16/samsung_files_patent_case_against_apple_in_australia_over_iphone_ipad.html). Appleinsider.com. . Retrieved 2012-01-05.
[157] "Microsoft and Samsung Broaden Smartphone Partnership: Agreements mark new initiatives to promote Windows Phone and share intellectual property" (http://www.microsoft.com/Presspass/press/2011/sep11/09-28SamsungPR.mspx). Microsoft.com. 2011-09-28. . Retrieved 2012-01-05.
[158] Our Licensing Deal with Samsung: How IP Drives Innovation and Collaboration - Microsoft on the Issues - Site Home - TechNet Blogs (http://blogs.technet.com/b/microsoft_on_the_issues/archive/2011/09/28/our-licensing-deal-with-samsung-how-ip-drives-innovation-and-collaboration.aspx).
[159] Florian Mueller (2011-09-28). "FOSS Patents: Samsung takes Android patent license from Microsoft rather than wait for Motorola" (http://fosspatents.blogspot.com/2011/09/samsung-takes-android-patent-license.html). Fosspatents.blogspot.com. . Retrieved 2012-01-05.
[160] "Apple Wins Injunction Blocking Sale of Galaxy Tab 10.1 in Australia" (http://www.macrumors.com/2011/10/12/apple-wins-injunction-blocking-sale-of-galaxy-tab-10-1-in-australia/). Mac Rumors. 2011-10-12. . Retrieved 2012-01-05.
[161] "Microsoft and Quanta Computer Sign Patent Agreement Covering Android and Chrome-Based Devices - Redmond, Wash., Oct. 13, 2011 /PRNewswire/" (http://www.prnewswire.com/news-releases/microsoft-and-quanta-computer-sign-patent-agreement-covering-android-and-chrome-based-devices-131793888.html). Washington: Prnewswire.com. . Retrieved 2012-01-05.
[162] Microsoft Inks Another Android Patent Deal, This Time With Quanta | TechCrunch (http://techcrunch.com/2011/10/13/microsoft-inks-another-android-patent-deal-this-time-with-quanta/).
[163] Levine, Dan. "U.S. judge says Samsung tablets infringe Apple patents" (http://www.reuters.com/article/2011/10/13/us-apple-samsung-lawsuit-idUSTRE79C79C20111013?feedType=RSS&feedName=businessNews&utm_source=dlvr.it&utm_medium=twitter&dlvrit=56943). Reuters.com. . Retrieved 2012-01-05.
[164] Apple Inc.. "Distribute your App - iOS Developer Program - Apple Developer" (http://developer.apple.com/programs/ios/distribute.html). Developer.apple.com. . Retrieved 2012-01-05.
[165] "Apple's rivals battle for iOS scraps as app market sales grow to $2.2 billion" (http://www.appleinsider.com/articles/11/02/18/rim_nokia_and_googles_android_battle_for_apples_ios_scraps_as_app_market_sales_grow_to_2_2_billion.html). Appleinsider.com. 2011-02-18. . Retrieved 2012-01-05.
[166] "Google Android has double the number of free apps than Apple's App Store" (http://techcrunch.com/2010/07/05/distimo-june-2010/). Distimo. 15 July 2009. . Retrieved 15 July 2009.
[167] "McAfee: Android malware surges 76%, iPhone untouched" (http://www.electronista.com/articles/11/08/23/mcafee.shows.android.facing.huge.spike.in.malware/). Electronista.com. . Retrieved 2012-01-05.

[168] Dalrymple, Jim (2011-11-16). "Android sees a 472% increase in malware since July" (http://www.loopinsight.com/2011/11/16/android-sees-a-472-increase-in-malware-since-july/). Loopinsight.com. . Retrieved 2012-01-05.

[169] Mobile Malware Development Continues To Rise, Android Leads The Way (http://globalthreatcenter.com/?p=2492).

[170] "The Mother Of All Android Malware Has Arrived" (http://www.androidpolice.com/2011/03/01/the-mother-of-all-android-malware-has-arrived-stolen-apps-released-to-the-market-that-root-your-phone-steal-your-data-and-open-backdoor/). *Android Police*. March 6, 2011. .

[171] Perez, Sarah (2009-02-12). "Android Vulnerability So Dangerous, Owners Warned Not to Use Phone's Web Browser" (http://www.readwriteweb.com/archives/android_vulnerability_so_dangerous_shouldnt_use_web_browser.php). Readwriteweb.com. . Retrieved 2011-08-08.

[172] "Lookout, Retrevo warn of growing Android malware epidemic, note Apple's iOS is far safer" (http://www.appleinsider.com/articles/11/08/03/lookout_retrevio_warn_of_growing_android_malware_epidemic_note_apples_ios_is_far_safer.html). Appleinsider.com. 2011-08-03. . Retrieved 2012-01-05.

[173] "First SMS Trojan detected for smartphones running Android" (http://www.kaspersky.com/news?id=207576158). Kaspersky Lab. . Retrieved 2010-10-18.

[174] "Apple's iOS unaffected by malware as Android exploits surge 76%" (http://www.appleinsider.com/articles/11/08/24/apples_ios_unaffected_by_malware_as_android_exploits_surge_76.html). Appleinsider.com. 2011-08-24. . Retrieved 2012-01-05.

[175] "Security researcher finds code signing flaw that opens door for iOS malware" (http://www.appleinsider.com/articles/11/11/07/newly_found_code_signing_flaw_allows_for_ios_malware.html). Appleinsider.com. 2011-11-07. . Retrieved 2012-01-05.

[176] Apple's iOS more secure than Google's Android, says Symantec (http://www.appleinsider.com/articles/11/06/28/apples_ios_more_secure_than_googles_android_says_symantec.html)

[177] the understatement: Android Orphans: Visualizing a Sad History of Support (http://theunderstatement.com/post/11982112928/android-orphans-visualizing-a-sad-history-of-support)

[178] "Smart phones: how to stay clever in downturn" (http://www.deloitte.co.uk/TMTPredictions/telecommunications/Smartphones-clever-in-downturn.cfm). *Deloitte Telecommunications Predictions*. .

[179] "Android Phones Steal Market Share" (http://bmighty.informationweek.com/mobile/showArticle.jhtml?articleID=224201881). .

[180] "100 Million Club – H1 2010" (http://www.visionmobile.com/blog/2010/10/smart-feature-phones-the-unbalanced-equation-100-million-club-series/). .

[181] "Mobile Phone Development » Blog Archive » Gartner Q3 2010" (http://www.mobilephonedevelopment.com/archives/1149). Mobilephonedevelopment.com. 2010-11-10. . Retrieved 2011-09-07.

[182] Stan Schroeder (2011-02-10). "Gartner: Symbian Is Still the Number One Smartphone Platform" (http://mashable.com/2011/02/10/symbian-number-one-gartner-report/). Mashable.com. . Retrieved 2011-09-07.

[183] "Gartner Says Worldwide Mobile Device Sales to End Users Reached 1.6 Billion Units in 2010; Smartphone Sales Grew 72 Percent in 2010" (http://www.gartner.com/it/page.jsp?id=1543014). Gartner.com. 2011-02-11. . Retrieved 2011-09-07.

[184] "Olswang Predicts Mobile's Impact on TV: The Convergent Revolution" (http://www.tvgenius.net/blog/2011/03/17/mobile-tv-convergence/). Tvgenius.net. 2011-03-17. . Retrieved 2011-09-07.

[185] "Berg: Smartphone shipments grew 74% in 2010" (http://www.bgr.com/2011/03/10/berg-smartphone-shipments-grew-74-in-2010/). *Boy Genius Report*. March 10, 2011. .

[186] Generation App: 62% of Mobile Users 25-34 own Smartphones | Nielsen Wire (http://blog.nielsen.com/nielsenwire/?p=29786).

[187] Market Research | Consumer Market Research - NPD - As Smartphone Prices Fall, Retailers Are Leaving Money on the Table, According to The NPD Group (http://www.npdgroup.com/wps/portal/npd/us/news/pressreleases/pr_111114a)

[188] Apple, With 4 Percent of Handset Market, Captures 52 Percent of Profits (http://www.pcmag.com/article2/0,2817,2395951,00.asp#fbid=SLgcdJun7IU)

[189] "Smartphones killing point-and-shoots, now take almost 1/3 of photos" (http://gigaom.com/2011/12/22/smartphones-killing-point-and-shoots-now-take-almost-13-of-photos/). . Retrieved December 25, 2011.

[190] "Gartner Says Sales of Mobile Devices in Second Quarter of 2011 Grew 16.5 Percent Year-on-Year; Smartphone Sales Grew 74 Percent" (http://www.gartner.com/it/page.jsp?id=1764714). Gartner. 11 August 2011. . Retrieved 20 August 2011.

[191] Tarmo Virki (2011-01-31). "Google topples Symbian from smart phones top spot" (http://www.theglobeandmail.com/news/technology/mobile-technology/google-topples-symbian-from-smartphones-top-spot/article1888557/?cmpid=rss1). Theglobeandmail.com. . Retrieved 2011-09-07.

[192] "Android became clear smartphone leader in first quarter: Canalys" (http://www.reuters.com/article/2011/05/04/us-smartphones-research-idUSTRE7434VV20110504). Reuters.com. 2011-05-04. . Retrieved 2011-09-07.

[193] "Android Takes Over in UK" (http://www.bestsmartphone.com/2011/11/02/android-takes-over-in-uk/). bestsmartphone.com. 2011-11-02. .

[194] http://www.gartner.com/it/page.jsp?id=910112

[195] http://www.gartner.com/it/page.jsp?id=910112

[196] http://www.gartner.com/it/page.jsp?id=1306513

[197] http://www.gartner.com/it/page.jsp?id=1543014

[198] iPass : Mobile Workforce Report (http://mobile-workforce-project.ipass.com/).

[199] iPhone pushes past BlackBerry to top enterprise phone ranks (http://www.appleinsider.com/articles/11/11/16/iphone_pushes_past_blackberry_to_top_enterprise_phone_ranks.html)

[200] In the US Market, iPhone Outperforms Other Mobile Platforms in User Loyalty by a Wide Margin, Android is Second, Blackberry Fourth (http://www.zokem.com/2011/01/in-the-us-market-iphone-outperforms-other-mobile-platforms-in-user-loyalty-by-a-wide-margin-android-is-second-blackberry-fourth/)

[201] iPhone Whips Android, BlackBerry in User Loyalty: Zokem (http://www.eweek.com/c/a/Mobile-and-Wireless/iPhone-Whips-Android-Blackberry-in-User-Loyalty-Zokem-445681/)

[202] Strategy Analytics: Samsung Becomes World's Number One Smartphone Vendor in Q3 2011 - MarketWatch (http://www.marketwatch.com/story/strategy-analytics-samsung-becomes-worlds-number-one-smartphone-vendor-in-q3-2011-2011-10-27)

[203] http://www.gartner.com/it/page.jsp?id=1764714

[204] "Smartphone Sales Will Hit 420 Million In 2011, To Take 28 Percent Of The Total Phone Market" (http://techcrunch.com/2011/07/27/smartphone-sales-will-hit-420-million-in-2011-to-take-28-percent-of-the-total-phone-market/). TechCrunch. 27 July 2011. .

[205] Apple's iPhone accounted for 66% of Q2 smartphone profit among top vendors (http://www.bgr.com/2011/07/29/apples-iphone-accounted-for-66-of-q2-smartphone-profit-among-top-vendors/)

[206] "SA agrees: Apple now top smartphone vendor in the world with 140% growth" (http://www.bgr.com/2011/07/29/sa-agrees-apple-now-top-smartphone-vendor-in-the-world-with-240-growth/). BGR. 29 July 2011. .

[207] "Nokia reports Q2 results: Sells 88.5 million devices and declares net loss of 368 million euros" (http://mobilesyrup.com/2011/07/21/nokia-reports-q2-results-sells-88-5-million-devices-and-declares-net-loss-of-368-million-euros/). Mobilesyrup.com. 2011-07-21. . Retrieved 2011-09-07.

[208] "In U.S. Smartphone Market, Android is Top Operating System, Apple is Top Manufacturer" (http://blog.nielsen.com/nielsenwire/?p=28516). Nielsen. 28 July 2011. . Retrieved 5 September 2011.

[209] "Android Still Dominates Phones, But What About the Rest of Mobile?" (http://www.wired.com/gadgetlab/2011/07/android-ios-platform-share/all/1). Wired. 28 July 2011. . Retrieved 5 September 2011.

[210] iPhone 4 remains top-selling US smartphone despite growing iPhone 5 hype (http://www.appleinsider.com/articles/11/09/06/iphone_4_remains_top_selling_us_smartphone_despite_growing_iphone_5_hype.html)

[211] Previous-gen Apple iPad, iPhone 3GS often outsell new Android devices (http://www.appleinsider.com/articles/11/05/09/previous_gen_apple_ipad_iphone_3gs_often_outsell_new_android_devices.html)

[212] "Samsung Becomes Biggest Smartphone Vendor, as Android's Market Share Grows" (http://www.pcworld.com/article/243861/samsung_becomes_biggest_smartphone_vendor_as_androids_market_share_grows.html). PCWorld. 2011-11-15. . Retrieved 2011-12-10.

[213] Ever-Popular iPhone Named Top Smartphone. Again (http://www.wired.com/gadgetlab/2011/09/iphone-tops-survey-sixth-time/all/1).

[214] Wireless Consumer Smartphone Ratings (Volume 1) (http://www.jdpower.com/Electronics/ratings/wireless-consumer-smartphone-ratings-(volume-1)/)

[215] Wireless Consumer Smartphone Ratings (Volume 2) (http://www.jdpower.com/Electronics/ratings/wireless-consumer-smartphone-ratings-(volume-2)/).

[216] 2011 U.S. Wireless Handset Customer Satisfaction Studies—Vol. 2 (http://www.jdpower.com/news/pressRelease.aspx?ID=2011146)

[217] 2011 Wireless Smartphone and Traditional Mobile Phone Satisfaction Studies-Vol. 1 (http://www.jdpower.com/news/pressrelease.aspx?ID=2011030)

[218] 2010 U.S. Wireless Smartphone and Traditional Mobile Phone Customer Satisfaction Studies-Volume 1 (http://businesscenter.jdpower.com/news/pressrelease.aspx?ID=2010039)

[219] 2009 Wireless Consumer Smartphone and Traditional Mobile Phone Customer Satisfaction Studies (http://businesscenter.jdpower.com/news/pressrelease.aspx?ID=2009082)

[220] Menezes, Gary (2010-09-11). "Symbian OS, Now Fully Open Source" (http://www.watblog.com/2010/02/06/symbian-os-now-fully-open-source). Watblog.com. . Retrieved 2011-09-07.

[221] "Nokia N9 Smart Phone Review" (http://wontek.com/smart-phones/nokia/nokia-n9-smart-phone-review). Wontek.com. 2011-01-01. . Retrieved 2011-09-07.

[222] Lance Whitney, CNET. " Nearly 1 in 5 smartphone owners use check-in services (http://news.cnet.com/8301-1023_3-20062640-93.html?part=rss&subj=news&tag=2547-1_3-0-20)." May 13, 2011. Retrieved May 13, 2011.

[223] Second Screen change the way we watch TV (http://www.good.is/post/double-the-glow-will-second-screen-apps-change-the-way-we-watch-tv/)

[224] explosion of second screen apps (http://www.adweek.com/news/technology/second-screen-apps-explode-132237)

External links

- Freesmartphone.org (http://www.freesmartphone.org) – a collaboration platform for open-source smartphones
- Defining the Smartphone, part 1 (http://www.allaboutsymbian.com/features/item/Defining_the_Smartphone.php), part 2 (http://www.allaboutsymbian.com/features/item/Spy_versus_Spy_No_Smartphone_versus_Smartphone.php) by Steve Litchfield on July 16, 2010, various definitions are treated
- Mozilla Labs: Seabird –Community driven Mobile Phone Concept (http://mozillalabs.com/conceptseries/2010/09/23/seabird/)

Tablet_computer

A **tablet computer**, or a **tablet**, is a mobile computer, larger than a mobile phone or personal digital assistant, integrated into a flat touch screen and primarily operated by touching the screen rather than using a physical keyboard. It often uses an onscreen virtual keyboard, a passive stylus pen, or a digital pen.[1] [2]

The Apple iPad

The term may also apply to a variety of form factors that differ in position of the screen with respect to a keyboard. The standard form is called *slate*, which does not have an integrated keyboard but may be connected to one with a wireless link or a USB port. *Convertible* notebook computers have an integrated keyboard that can be hidden by a swivel joint or slide joint, exposing only the screen for touch operation. *Hybrids* have a detachable keyboard so that the touch screen can be used as a stand-alone tablet. *Booklets* include two touch screens, and can be used as a notebook by displaying a virtual keyboard in one of them.

Early examples of the *information tablet* concept originated in the 19th and 20th centuries mainly as prototypes and concept ideas, the more prominent the Alan Kay's Dynabook. First commercial portable electronic devices based on the concept appeared at the end of the 20th century. During the 2000s Microsoft attempted a relatively unsuccessful product line with Microsoft Tablet PC, which carved a niche market at hospitals and outdoor businesses. In 2010 Apple released the iPad based on the technology developed in parallel with their previous iPhone, and reached worldwide commercial success.

Background

History

The tablet computer and the associated special operating software is an example of pen computing technology, and thus the development of tablets has deep historical roots.

Electrical devices with data input and output on a flat information display have existed as early as 1888.[3] Throughout the 20th century many devices with these characteristics have been ideated and created whether as blueprints, prototypes or commercial products, with the Dynabook concept in 1968 being a spiritual precursor of tablets and laptops. In addition to many academic and research systems, there were several companies with commercial products in the 1980s.

During the 2000s Microsoft attempted to define with the Microsoft Tablet PC the *tablet personal computer* product concept[4] as a mobile computer for field work in business,[5] though their devices failed to achieve widespread usage mainly due to price and usability problems that made them unsuitable outside of their limited intended purpose.[6]

In April 2010 Apple Inc. released the iPad, a tablet computer with an emphasis on media consumption. The shift in purpose, together with increased usability, battery life, simplicity, lower weight and cost, and overall quality with respect to previous tablets, was perceived as defining a new class of consumer device[7] and shaped the commercial market for tablets in the following year.[8]

As a result, two distinctly different types of tablet computing devices exist as of 2011, the *Tablet PC* and the *Post-PC tablet*, whose operating systems are of different origin.

Traditional tablet PCs

A tablet personal computer (tablet PC) is a portable personal computer equipped with a touchscreen as a primary input device, and running a modified desktop OS[9] designed to be operated and owned by an individual.[10] The term was made popular as a concept presented by Microsoft in 2000[11] and 2001[12] but tablet PCs now refer to any tablet-sized personal computer regardless of the (desktop) operating system.[13]

Tablet personal computers are mainly based on the x86 IBM-PC architecture [14] and are fully functional personal computers employing a slightly modified personal computer OS (such as Windows or Ubuntu Linux) supporting their touch-screen, instead of a traditional display, mouse and keyboard. A typical tablet personal computer needs to be stylus driven, because operating the typical desktop based OS requires a high precision to select GUI widgets, such as a the close window button.

"Post-PC" tablets

Since mid-2010, new tablet computers have been introduced with mobile operating systems that forgo the Wintel paradigm,[15] have a different interface instead of the traditional desktop OS, and represent a new type of computing device.[16] These "post-PC" mobile OS tablet computer devices are normally finger driven and most frequently use capacitive touch screens with multi-touch capabilities instead of the simple resistive touchscreens of typical stylus driven systems.

The most successful of these was the Apple iPad using the iOS operating system.[17] Samsung's Galaxy Tab and others followed, continuing the now common trends towards multi-touch and other natural user interface features, as well as flash memory solid-state storage drives and "instant on" warm-boot times; in addition, standard external USB and Bluetooth keyboards can often be used. Most frequently the operating system running a tablet computer that's not based on the traditional PC architecture is based on a Unix-like OS, such as Darwin, Linux or QNX. Some have 3G mobile telephony capabilities.[18]

In forgoing the x86 precondition (a requisite of Windows compatibility), most tablet computers released since mid-2010 use a version of an ARM architecture processor for longer battery life versus battery weight, heretofore used in portable equipment such as MP3 players and cell phones. Especially with the introduction of the ARM Cortex family, this architecture is now powerful enough for tasks such as internet browsing, light production work and gaming.[19]

A significant trait of tablet computers not based on the traditional PC architecture is that the main source of 3rd party software for these devices tends to be through online distribution, rather than more traditional methods of boxed software or direct sales from software vendors. These sources, known as "app stores," provide centralized catalogues of software from both 1st and 3rd parties and allow simple "one click" on-device software purchasing, installation, and updates.

Touch user interface

Samsung Galaxy Tab demonstrating multi-touch

A key and common component among tablet computers is touch input. This allows the user to navigate easily and intuitively and type with a virtual keyboard on the screen.

The event processing of the operating system must respond to touches rather than clicks of a keyboard or mouse, which allows integrated hand-eye operation, a natural part of the somatosensory system. Although the device implementation differs from more traditional PCs or laptops, tablets are disrupting the current vendor sales by weakening traditional laptop PC sales in favor of the current tablet computers.[20] [21] [22] This is even more true of the "finger driven multi-touch" interface of the more recent tablet computers, which often emulate the way actual objects behave.

Handwriting recognition

Because tablet personal computers normally use a stylus, they quite often implement handwriting recognition, while other tablet computers with finger driven screens do not. Finger driven screens however are potentially better suited for inputting "variable width stroke based" characters, like Chinese/Japanese/Korean writing, due to their built in capability of "pressure sensing". However at the moment not much of this potential is already used, and as a result even on tablet computers Chinese users often use a (virtual) keyboard for input.[23]

Chinese characters like this one meaning "person" can be written by handwriting recognition (人, Mandarin: *rén*, Korean: *in*, Japanese: *jin*, *nin*; *hito*, Cantonese: jan4). The character has two strokes, the first shown here in dark, and the second in red. The black area represents the starting position of the writing instrument.

Touchscreen hardware

Touchscreens are usually one of two forms;

- Resistive touchscreens are passive and can respond to any kind of pressure on the screen. They allow a high level of precision (which may be needed, when the touch screen tries to emulate a pointer for precision pointing, which in Tablet personal computers is common) but may require calibration to be accurate. Because of the high resolution of detection, a stylus or fingernail is often used for resistive screens. Although some possibility exist for implementing multi-touch on a resistive touch-screen, the possibilities are quite limited. As modern tablet computers tend to heavily lean on the use of multi-touch, this technology has faded out on high-end devices where it has been replaced by capacitive touchscreens.
- Capacitive touchscreens tend to be less accurate, but more responsive than resistive screens. Because they require a conductive material, such as a finger tip, for input, they are not common among (stylus using) Tablet PCs but are more prominent on the smaller scale "tablet computer" devices for ease of use, which generally do not use a stylus, and need multi-touch capabilities.

Other touch technology used in tablets include:

- Palm recognition. It prevents inadvertent palms or other contacts from disrupting the pen's input.
- Multi-touch capabilities, which can recognize multiple simultaneous finger touches, allowing for enhanced manipulation of on-screen objects.[24]

Some professional-grade Tablet PCs use pressure sensitive films that additionally allows pressure sensitivity such as those on graphics tablets.

Concurrently capacitive touch-screens, which use finger tip detection can often detect the size of the touched area, and can make some conclusions to the pressure force used, for a similar result.[25]

Other features

- Accelerometer: An accelerometer is a device that detects the physical movements of the tablet. This allows greater flexibility of use since tablets do not necessarily have a fixed direction of use. The accelerometer can also be used to detect the orientation of the tablet relative to the center of the earth, but can also detect movement of the tablet, both of which can be used as an alternative control interface for a tablet's software.
- Ambient light and proximity sensors are additional "senses", that can provide controlling input for the tablet.
- Storage drive: Large tablets use storage drives similar to laptops, while smaller ones tend to use drives similar to MP3 players or have on-board flash memory. They also often have ports for removable storage such as Secure Digital cards. Due to the nature of the use of tablets, solid-state memory is often preferable due to its better resistance to damage during movement.
- Wireless: Because tablets by design are mobile computers, wireless connections are less restrictive to motion than wired connections. Wi-Fi connectivity has become ubiquitous among tablets. Bluetooth is commonly used for connecting peripherals and communicating with local devices in place of a wired USB connection.
- 3D: Following mobile phone, there are also 3D slate tablet with dual lens at the back of the tablet and also provided with blue-red glasses.[26]
- Docking station: Some newer tablets are offering a optional docking station that has a full size qwerty keyboard and USB port, providing both portability and flexibility.

Form factors

Tablet computers come in a range of sizes, currently ranging from tablet PCs to PDAs. Tablet personal computers tend to be as large as laptops and often are the largest usable size for mobile tablet computing while the new generation of tablet computers can be (much) smaller and use a RISC (ARM or MIPS) CPU, and in size can border on PDAs.

Slate

Slate computers, which resemble writing slates, are tablet computers without a dedicated keyboard. For text input, users rely on handwriting recognition via an active digitizer, touching an on-screen keyboard using fingertips or a stylus, or using an external keyboard that can usually be attached via a wireless or USB connection.

Slate computers typically incorporate small (8.4–14.1 inches/21–36 centimetres) LCD screens and have been popular in vertical markets such as health care, education, hospitality, aviation (pilot documentation and maps),[27] and field work. Applications for field work often require a tablet computer that has rugged specifications that ensure long life by resisting heat, humidity, and drop/vibration damage. This added focus on mobility and/or ruggedness often leads to elimination of moving parts that could hinder these qualities.

Booklet

Booklet computers are dual-touchscreen tablet computers that fold like a book. Typical booklet computers are equipped with multi-touch screens and pen writing recognition capabilities. They are designed to be used as digital day planners, Internet surfing devices, project planners, music players, and displays for video, live TV, and e-reading.

Convertible

Convertible notebooks have a base body with an attached keyboard. They more closely resemble modern laptops, and are usually heavier and larger than slates.

Typically, the base of a convertible attaches to the display at a single joint called a swivel hinge or rotating hinge. The joint allows the screen to rotate through 180° and fold down on top of the keyboard to provide a flat writing surface. This design, although the most common, creates a physical point of weakness on the notebook.

A Lenovo X61 in slate mode

Some manufacturers have attempted to overcome these weak points. The Panasonic Toughbook 19, for example, is advertised as a more durable convertible notebook. One model by Acer (the TravelMate C210) has a sliding design in which the screen slides up from the slate-like position and locks into place to provide the laptop mode.

Sliding screens were presented at CES 2011. The first product to use it is the Samsung Sliding PC7 Series,[28] a tablet with Intel Atom hardware and a unique sliding screen that allows the product to be used as a laptop or slate tablet when the screen is locked in place covering the whole keyboard. The concept still has to prove its reliability, but is intended to combine the virtues of tablet PCs with those of notebooks. Also presented was the upcoming Inspiron Duo from Dell, which rotates the screen horizontally when opened. Convertibles like that with hardware specs of a netbook are called netvertibles.

Hybrid

Hybrids, a term coined by users of the HP/Compaq TC1000 and TC1100 series, share the features of the slate and convertible by using a detachable keyboard that operates in a similar fashion to a convertible when attached. Hybrids are not to be confused with slate models with detachable keyboards; detachable keyboards for pure slate models do not rotate to allow the tablet to rest on it like a convertible.

System architecture

Two major computer architectures compete in the tablet market,[29] x86 and ARM architecture. x86, including x86-64, is popular on tablet PCs due to its use on laptops which can share common software and hardware and which can run a version of Windows. There are also non-PC based x86 tablets like the JooJoo. ARM gained popularity following the success of the iPad.[30] ARM is more power and cost efficient for mobile computing and is gaining popularity for smaller tablets from other manufacturers such as Samsung with the Galaxy Tab which runs on Android.

Operating systems and vendors

Tablets, like regular computers, can run a number of operating systems. These come in two classes, namely traditional desktop-based operating systems and post-PC mobile-based ("phone-like") operating systems.

For the former class popular OS's are Microsoft Windows, and a range of Linux distributions. HP is developing enterprise-level tablets under Windows and consumer-oriented tablets under webOS. In the latter class the popular variants include Apple iOS, and Google Android. Manufacturers are also testing the market for products with Windows CE, Chrome OS,[31] [32] and so forth.

Boot times for iPads are one-half the boot times for current Windows 7 netbooks, which can take over 50 seconds to display the login prompt.[33] The BIOS initialization for a PC, which has remained unchanged since the invention of the PC, can still take 25 seconds.[34]

Traditional Tablet PC operating systems

Microsoft

Following Windows for Pen Computing, Microsoft has been developing support for tablets runnings Windows under the Microsoft Tablet PC name.[35] According to a 2001 Microsoft definition[36] of the term, "Microsoft Tablet PCs" are pen-based, fully functional x86 PCs with handwriting and voice recognition functionality. Tablet PCs use the same hardware as normal laptops but add support for pen input. For specialized support for pen input, Microsoft released Windows XP Tablet PC Edition. Today there is no tablet specific version of Windows but instead support is built in to both Home and Business versions of Windows Vista and Windows 7. Tablets running Windows get the added functionality of using the touchscreen for mouse input, hand writing recognition, and gesture support. Following Tablet PC, Microsoft announced the UMPC initiative in 2006 which brought Windows tablets to a smaller, touch-centric form factor. This was relaunched in 2010 as *Slate PC*, to promote tablets running Windows 7, ahead of Apple's iPad launch.[37] [38] Slate PCs are expected to benefit from mobile hardware advances derived from the success of the netbooks.

Microsoft has since announced Windows 8 which will have features designed for touch input, while running on both PCs and ARM architecture.[39] Renee James of Intel states multiple builds are needed, with 1 build for x86 processors and with 4 builds for ARM; ARM targets are defined for AMD, Intel, Qualcomm, and TI processors.[40]

While many tablet manufacturers are moving to the ARM architecture with lighter operating systems, Microsoft has stood firm to Windows.[41] [42] [43] [44] Though Microsoft has Windows CE for ARM support it has kept its target market for the smartphone industry with Windows Mobile and the new Windows CE 6 based Windows Phone. Some manufacturers, however, still have shown prototypes of Windows CE-based tablets running a custom shell.[45] To date, the full Windows 7 does not yet support ARM architecture.[46]

Linux

One early implementation of a Linux tablet was the ProGear by FrontPath. The ProGear used a Transmeta chip and a resistive digitizer. The ProGear initially came with a version of Slackware Linux, but could later be bought with Windows 98. Because these computers are general purpose IBM PC compatible machines, they can run many different operating systems. However, the device is no longer for sale and FrontPath has ceased operations. It is important to note that many touch screen sub-notebook computers can run any of several Linux distributions with little customization.

X.org now supports screen rotation and tablet input through Wacom drivers, and handwriting recognition software from both the Qt-based Qtopia and GTK+-based Internet Tablet OS provide promising free and open source systems for future development. KDE's Plasma Active is graphical environments for tablet.[47]

Open source note taking software in Linux includes applications such as Xournal (which supports PDF file annotation), Gournal (a Gnome based note taking application), and the Java-based Jarnal (which supports handwriting recognition as a built-in function). Before the advent of the aforementioned software, many users had to rely on on-screen keyboards and alternative text input methods like Dasher. There is a stand alone handwriting recognition program available, CellWriter, which requires users to write letters separately in a grid.

A number of Linux based OS projects are dedicated to tablet PCs, but many desktop distributions now have tablet-friendly interfaces allowing the full set of desktop features on the smaller devices. Since all these are open source, they are freely available and can be run or ported to devices that conform to the tablet PC design. Maemo (rebranded MeeGo in 2010), a Debian Linux based graphical user environment, was developed for the Nokia Internet Tablet devices (770, N800, N810 & N900). It is currently in generation 5, and has a vast array of applications available in both official and user supported repositories. Ubuntu since version 11.04 has used the tablet-friendly Unity UI, and many other distributions (such as Fedora) use the also tablet-friendly Gnome shell (which can also be installed in Ubuntu if preferred). Previously the Ubuntu Netbook Remix edition was one of the only linux distibutions offering a tablet interface with all the applications and features of a desktop distribution, but this has been phased out with the expansion of Unity to the desktop. A large number of distributions now have touchscreen support of some kind, even if their interfaces are not well suited to touch operation.

Canonical has hinted that Ubuntu will be availiable on tablets, as well as phones and smart televisions, by 2014.[48]

TabletKiosk currently offers a hybrid digitizer / touch device running openSUSE Linux. It is the first device with this feature to support Linux.

Intel and Nokia

Nokia entered the tablet space with the Nokia 770 running Maemo, a Debian-based Linux distribution custom-made for their Internet tablet line. The product line continued with the N900 which is the first to add phone capabilities. The user interface and application framework layer, named Hildon, was an early instance of a software platform for generic computing in a tablet device intended for internet consumption.[49] But Nokia didn't commit to it as their only platform for their future mobile devices and the project competed against other in-house platforms. The strategic advantage of a modern platform was not exploited, being displaced by the Series 60. [50]

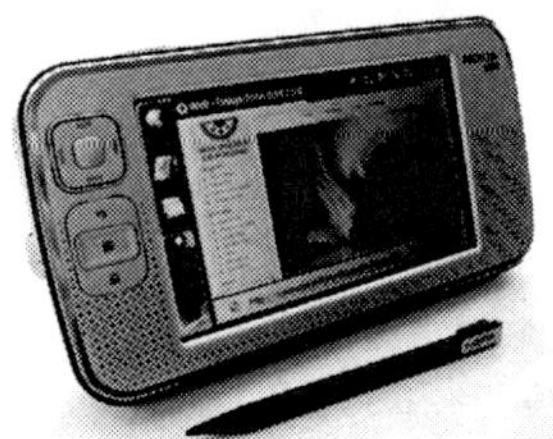

The Nokia N800

Intel, following the launch of the UMPC, started the Mobile Internet Device initiative, which took the same hardware and combined it with a Linux operating system custom-built for portable tablets. Intel co-developed the lightweight Moblin operating system following the successful launch of the Atom CPU series on netbooks. Intel is also setting tablet goals for Atom, going forward from 2010.[51] [52]

MeeGo

MeeGo is a Linux-based operating system developed by Intel and Nokia that supports Netbooks, Smartphones and Tablet PCs. In 2010, Nokia and Intel combined the Maemo and Moblin projects to form MeeGo. The first tablet using MeeGo is the Neofonie WeTab launched September 2010 in Germany. The WeTab uses an extended version of the MeeGo operating system called WeTab OS. WeTab OS adds runtimes for Android and Adobe AIR and provides a proprietary user interface optimized for the WeTab device. On 27 September 2011 it was announced by the Linux Foundation that MeeGo will be replaced in 2012 by Tizen, an open source mobile operating system.[53]

Post-PC operating systems

Tablets not following the personal computer (PC) tradition use operating systems in the style of those developed for PDAs and smartphones.

iPad

The iPad in case

The iPad runs a version of iOS which was first created for the iPhone and iPod Touch. Although built on the same underlying Unix implementation as MacOS, the operating system differs radically at the graphical user interface level. iOS is designed for finger based use and has none of the tiny features which required a stylus on earlier tablets. Apple introduced responsive multi touch gestures, like moving two fingers apart to zoom in. iOS is built for the ARM architecture, which uses less power, and so gives better battery life than the Intel devices used by Windows tablets. Previous to the iPad's launch, there were long standing rumors of an Apple tablet, though they were often about a product running Mac OS X and being in line with Apple's Macintosh computers.[54] This became partially true when a 3rd party offered customized Macbooks with pen input, known as the Modbook.

Previous to Apple's commercialization of the iPad, Axiotron introduced at Macworld in 2007[55] an aftermarket, heavily modified Apple MacBook called Modbook, a Mac OS X-based tablet personal computer. The Modbook uses Apple's Inkwell for handwriting and gesture recognition, and uses digitization hardware from Wacom. To get Mac OS X to talk to the digitizer on the integrated tablet, the Modbook is supplied with a third-party driver called TabletMagic [56]; Wacom does not provide driver support for this device.

Blackberry

The BlackBerry PlayBook is a tablet computer announced in September 2010 which runs the BlackBerry Tablet OS.[57] The OS is based on the QNX system that Research in Motion acquired in early 2010. Delivery to developers and enterprise customers is expected in October 2010. The BlackBerry PlayBook was officially released to US and Canadian consumers on April 19, 2011.

Android

An ASUS Eee Pad Transformer running Android 3.2.1 Honeycomb; the keyboard is part of a docking station for the tablet.

Google's Linux-based Android operating system has been targeted by tablet manufacturers following its success on smartphones due to its open nature and support for low-cost ARM systems much like Apple's iOS. In 2010, there have been numerous announcements of such tablets.[58] However, much of Android's tablet initiative comes from manufacturers as Google primarily focuses its development on smartphones and restricts the App Market from non-phone devices.[59] Toshiba's AC100 laptop also runs on Android.[60]

There is talk of tablet support from Google coming to its web-centric Chrome OS.[61] [62]

Some vendors such as Motorola[63] and Lenovo[64] are delaying deployment of their tablet computers until after 2011, after Android is reworked to include more tablet features.[65] Android 3.0 (Honeycomb) is optimized specifically for devices with larger screen sizes, mainly tablets, and has access to the Android Market. Android is the software stack for mobile devices that includes operating system, middleware and key applications.

HP

Hewlett Packard announced the TouchPad, running webOS 3.0 on a 1.2Ghz Snapdragon CPU, would be released in June 2011. On August 18, 2011, HP announced the discontinuation of the TouchPad, due to sluggish sales.[66] HP has announced that they will release webOS as open-source.[67]

One Laptop per Child organization

The One Laptop per Child (OLPC) organization is developing a new version of the OLPC, strongly resembling a tablet computer, called the OLPC XO-3, running its "Sugar" operating system, based on Linux. The new XO-3 will be based on ARM technology from Marvell.[68]

Developing software for tablet computers

The new class of devices heralded by the iPad has spurred the tendency of a walled garden approach where the vendor reserves rights as to what can be installed. The software development kits for these platforms are restricted and the vendor must approve the final application for distribution to users. These restrictions allow the hardware vendor to control the kind of software that can be used and the content that can be seen in the devices; this can be used to reduce the impact of malware on the platform and to provide material of approved content rating, and also to exclude software and content from competing vendors. The walled garden approach to application development has proven to be a competitive advantage for the iPad over HP's Touchpad, triggering HP's withdrawal from the industry, due in large part to sluggish Touchpad sales, after only 49 days on the market.[69]

Proponents of open source software deem that these restrictions on software installation and lack of administrator rights make this category one that, in their view, cannot be properly named "personal computers".[70] [71] [72] Some newer tablet computers using mobile operating systems don't use the walled garden concept, and are like personal computers in this regard.

Comparison with laptop computers

The advantages and disadvantages of tablet computers are highly subjective measures. What appeals to one user may be exactly what disappoints another. The following are commonly cited opinions of tablet computers versus laptops:

Advantages

- Usage in environments not conducive to a keyboard and mouse such as lying in bed, standing, or handling with a single hand.
- Lighter weight, lower power models can function similarly to dedicated E-book readers like the Amazon Kindle.
- Touch environment makes navigation easier than conventional use of keyboard and mouse or touch pad in certain contexts such as image manipulation, musical, or mouse oriented games.
- Digital painting and image editing are more precise and intuitive than painting or sketching with a mouse.
- The ability for easier or faster entry of diagrams, mathematical notations, and symbols.
- Allows, with the proper software, universal input, independent from different keyboard localizations.
- Some users find it more direct and pleasant to use a stylus, pen or finger to point and tap on objects, rather than use a mouse or touchpad, which are not directly connected to the pointer on screen.
- Current tablets typically have longer battery life than laptops or netbooks.

Disadvantages

- Higher price – convertible tablet computers can cost significantly more than non-tablet portable PCs although this premium has been predicted to fall.[73]
- Input speed – handwriting or typing on a virtual keyboard can be significantly slower than typing speed on a conventional keyboard, the latter of which can be as high as 50–150 WPM; however, Slideit, Swype and other technologies are offered in an effort to narrow the gap. Some devices also support external keyboards (eg: iPad can accept Bluetooth keyboards and USB keyboards through iPad Camera Connection Kit Dock Connector-to-USB adapter. Apple has also released a keyboard dock)
- Ergonomics – a tablet computer, or a folded slate PC, does not provide room for a wrist rest. In addition, the user will need to move his or her arm constantly while writing.
- More knowledge of the programs is needed – because information on e.g. icons isn't obtained by pointing at them. (The Compaq Concerto from 1992 didn't have this weakness.)
- Weaker video capabilities – Most tablet computers are equipped with embedded graphics processors instead of discrete graphics cards. In July 2010, the only tablet PC with a discrete graphics card was the HP TouchSmart tm2t, which has the ATI Mobility Radeon HD5450 as an optional extra.
- Business-oriented tablet personal computers have been slow sellers from 2001 to date.[74]
- Screen risk – Tablet computers are handled more than conventional laptops, yet many are built on similar frames; in addition, since their screens also serve as input devices, they run a higher risk of screen damage from impacts and misuse.
- Hinge risk – A convertible tablet computer's screen hinge is often required to rotate around two axes, unlike a normal laptop screen, subsequently increasing the number of possible mechanical and electrical (digitizer and video cables, embedded WiFi antennas, etc.) failure points.
- Smaller display and lack of keyboard.

Tablets in developing countries

The low hardware requirements and easy operation of tablet computers has made it subject to various design studies for use in developing countries, furthermore it will reduce the digital gap between the have and the have nots. Prototype tablet computers such as the Aakash have been projected to cost $35 with subsidy, according to researchers in India. It is expected to be available soon for the masses as the cheapest tablet working on Android with full functionality;[76] [77] however the bill of materials currently comes to $47 and will be available soon at retail shops at $60.[78] For the initial, the government will give them for free to students 100,000 of these tablets.[79]

OLPC XO-3, a tablet computer concept[75]

Before it, One laptop per child (OLPC) plans to introduce a tablet computer for $100.[80] Nicholas Negroponte, Chairman of OLPC, has invited the Indian researchers to MIT to begin sharing the OLPC design resources for their tablet computers.[81] OLPC has been awarded a grant for an interim step to their next generation tablet, OLPC XO-3.[82]

See also

- Comparison of tablet computers
- Smartbook
- Ultra-Mobile PC

References

[1] Editors PC Magazine. "Definition of: tablet computer" (http://www.pcmag.com/encyclopedia_term/0,2542,t=tablet+computer&i=52520,00.asp). *PC Magazine*. . Retrieved April 17, 2010.

[2] Editors Dictionary.com, "tablet computer – 1 dictionary result" (http://dictionary.reference.com/browse/tablet+computer), *Dictionary.com*, , retrieved April 17, 2010

[3] Gray, Elisha (1888-07-31), *Telautograph* (http://www.freepatentsonline.com/386815.pdf), United States Patent 386,815 (full image),

[4] John Markoff, *The New York Times*, August 30, 1999, " Microsoft brings in top talent to pursue old goal: the tablet (http://www.nytimes.com/1999/08/30/business/microsoft-brings-in-top-talent-to-pursue-old-goal-the-tablet.html)"

[5] "Tablet PC: Coming to an Office Near You?" (http://itmanagement.earthweb.com/netsys/article.php/1495701/Tablet-PC-Coming-to-an-Office-Near-You.htm). .

[6] Bright, Peter Ballmer (and Microsoft) still doesn't get the iPad (http://arstechnica.com/microsoft/news/2010/07/ballmer-and-microsoft-still-doesnt-get-the-ipad.ars), Ars Technica, 2010

[7] "The iPad's victory in defining the tablet: What it means" (http://www.infoworld.com/d/mobile-technology/the-ipads-victory-in-defining-the-tablet-what-it-means-431). Infoworld. .

[8] Gilbert, Jason (August 19, 2011). "HP TouchPad Bites The Dust: Can Any Tablet Dethrone The IPad?" (http://www.huffingtonpost.com/2011/08/19/hp-touchpad-ipad-tablet_n_931593.html). *Huffington Post*. .

[9] Beck H *et al*, *Business Communication and Technologies in a Changing World*, Macmillan Education Australia, 2009, p 402

[10] Haven, Kendall F. *100 greatest science inventions of all time*, Libraries Unlimited, 2006, p 191

[11] Bill Gates introduces Tablet PC, COMDEX Nov 2000 (http://www.microsoft.com/presspass/features/2000/nov00/11-13comdex.mspx)

[12] Page, M Microsoft Tablet PC Overview (http://www.transmetazone.com/articleview.cfm?articleID=499), TransmetaZone, December 21, 2000

[13] Kuhn, Bradley M. Free software and cellphones (http://www.fsf.org/working-together/next-steps/free-software-phones), Free Software Foundation, 2010

[14] Are Intel, AMD threatened by tablet growth? Every two to three tablets sold means one lost PC sale, analyst says (http://www.marketwatch.com/story/tablet-growth-may-threaten-pc-chip-makers-2010-10-22?pagenumber=2) accessdate=2010-10-24

[15] Roger Kay on Intel and Microsoft, as quoted April 29, 2011: "Clearly, each one is looking at a post-PC world..." MarketWatch (http://www.marketwatch.com/story/microsoft-offers-more-muted-view-of-pcs-2011-04-29)

[16] Lev Grossman (Thursday, Apr. 1, 2010) " Do We Need the iPad? A TIME Review (http://www.time.com/time/business/article/0,8599,1976932,00.html)", *TIME*

[17] Worstall, Tim. *Forbes*. http://www.forbes.com/sites/timworstall/2011/07/02/ipad-one-of-the-most-successful-products-ever/.

[18] i.e. ZTE V9 Tablet and Samsung Galaxy Tab and some iPads

[19] The Coming War: ARM versus x86 (http://vanshardware.com/2010/08/mirror-the-coming-war-arm-versus-x86/) Mirror for: *The Bright Side of News* April 8, 2010

[20] Best Buy: iPad cutting into laptop sales (http://news.cnet.com/8301-13579_3-20016818-37.html)

[21] Notebook sales growth goes negative. Can we blame the iPad yet? (http://tech.fortune.cnn.com/2010/09/17/notebook-sales-growth-goes-negative-can-we-blame-the-ipad-yet/)

[22] Tablets hurt PC sales but not Macs (http://www.marketwatch.com/story/tablets-hurt-pc-sales-but-not-apples-mac-2010-10-13)

[23] China using keyboards versus tablet input (http://webcache.googleusercontent.com/search?q=cache:rX3A5F0UJKQJ:ca.news.yahoo.com/s/afp/100826/technology/lifestyle_hongkong_china_japan_culture_technology+ca.news.yahoo.com/s/afp/100826/technology/lifestyle_hongkong_china_japan_culture_technology&cd=1&hl=nl&ct=clnk&gl=nl)

[24] jkOnTheRun:So what is multi-touch? (http://jkontherun.blogs.com/jkontherun/2007/12/so-what-is-mult.html)

[25] Buxton, Bill. "Multitouch Overview" (http://www.billbuxton.com/multitouchOverview.html)

[26] T-Mobile to sell tablet with 3-D cameras, glasses http://news.yahoo.com/s/ap/20110202/ap_on_hi_te/us_tec_techbit3_d_tablet

[27] *Los Angeles Times* (September 26, 2011 (http://www.latimes.com/business/la-fi-isoldiers-20110926,0,2255882,full.story))

[28] product presentation and demo Samsung Sliding PC7 Series (http://www.alltouchtablet.com/touchscreen-tablet-news/samsung-sliding-pc-7-is-the-slider-laptoptablet-you-ever-wanted-6343/), AllTouchTablet, 2011

[29] Intel has ARM in its crosshairs (http://news.cnet.com/Intel-has-ARM-in-its-crosshairs---page-2/2100-1006_3-6210033-2.html?tag=mncol)

[30] "Apple iPad Price, Features Say "ARM" All Over" (http://www.bnet.com/blog/mobile-internet/apple-ipad-price-features-say-8220arm-8221-all-over/133). bnet. .

[31] HP vice-president Todd Bradley projects HP Slates for enterprise-level tablets, webOS for consumer-level tablets accessdate=2010-10-5 (http://www.eweekeurope.co.uk/news/hp-hints-at-business-focused-windows-7-tablet-8704)
[32] HP Slate 500 runs Win 7 Pro (http://www.csmonitor.com/Innovation/Horizons/2010/1023/HP-Slate-500-brings-professional-spin-to-the-tablet-wars), an enterprise-level tablet from HP accessdate=2010-10-23
[33] Boot time comparisons for iPad vs netbook (http://reviews.cnet.com/8301-31747_7-20001653-243.html?tag=mncol;txt)
[34] Getting a Windows PC to boot in under 10 seconds (http://news.cnet.com/8301-13924_3-20018475-64.html#ixzz12IPFj4kf)
[35] Microsoft Tablet PC (http://msdn.microsoft.com/en-us/library/ms840465.aspx)
[36] Tablet PC Brings the Simplicity of Pen and Paper to Computing: In a conversation with PressPass, Tablet PC general manager Alexandra Loeb discusses how the Tablet PC will brin... (http://www.microsoft.com/presspass/features/2000/nov00/11-13tabletpc.mspx)
[37] "Live from Steve Ballmer's CES 2010 keynote" (http://www.engadget.com/2010/01/06/live-from-steve-ballmers-ces-2010-keynote/). Engadget. . Retrieved August 4, 2010.
[38] Initial Windows 7 tablets are slated to appear during holiday 2010 season. (http://news.cnet.com/8301-13860_3-20019267-56.html?tag=mncol;mlt_related) accessdate=2010-10-19
[39] Avi Greengart of Current Analysis states "Windows 8 basically assumes that every PC is a tablet", as reported by Gordon Mah Ung et.al. (October 2011) *Maximum PC* p.27 ISSN 1522-7249
[40] Gordon Mah Ung et.al. (October 2011) *Maximum PC* p.24 ISSN 1522-7249
[41] "Ballmer Admits Apple is Beating Microsoft in the Tablet Sector" (http://www.dailytech.com/Ballmer+Admits+Apple+is+Beating+Microsoft+in+the+Tablet+Sector/article19215.htm). DailyTech. . Retrieved August 6, 2010.
[42] Windows 7 is not yet optimized for fingertip events – September 24, 2010 (http://arstechnica.com/gadgets/news/2010/09/hp-slate-video-shows-all-thats-wrong-with-windows-7-on-tablets.ars)
[43] Windows 7 will not be optimized for slates; that will have to wait for Windows 8 (http://www.foxnews.com/scitech/2010/10/05/microsofts-ipad-answer-coming-christmas-holiday/)
[44] Windows 8 will not appear until 2012 (http://mashable.com/2010/10/24/microsoft-windows-8-2012/) accessdate=2010-10-24
[45] "Asus launches Eee Pad tablets and Eee Tablet note-taking thingie" (http://www.liliputing.com/2010/05/asus-launches-two-tablets-the-eee-pad-and-eee-tablet.html). liliputing. . Retrieved August 6, 2010.
[46] No Windows 7 for ARM (http://www.zdnet.com/blog/microsoft/microsoft-no-windows-7-for-arm-based-netbooks-for-now/2953?tag=mantle_skin;content) accessdate=2010-10-17

- Microsoft plans Windows tied to ARM chips (http://www.marketwatch.com/story/microsoft-plans-windows-tied-to-arm-chips-reports-2010-12-22?dist=beforebell) Dec. 22, 2010, 4:19 am EST

[47] "Plasma Active" (http://plasma-active.org/). .
[48] "Ubuntu coming to tablets, phones and smart TVs by 2014" (http://www.engadget.com/2011/10/31/ubuntu-coming-to-tablets-phones-cars-and-smart-tvs-by-2014/). .
[49] Andrew Orlowski. "Nokia's Great Lost Platform" (http://www.theregister.co.uk/2011/11/21/nokia_hildon_the_great_lost_platform/). The Register. .
[50] "Nokia's Great Lost Platform - Page 4" (http://www.theregister.co.uk/2011/11/21/nokia_hildon_the_great_lost_platform/page4.html).
.
[51] Intel shows pricing pressures for Atom due to competition from ARM (http://www.theregister.co.uk/2010/10/17/intel_tablets/) accessdate=2010-10-17
[52] Intel launches FPGA-equipped Atom (http://www.thinq.co.uk/2010/11/22/intel-launches-fpga-equipped-atom/) accessdate=2010-11-23 An FPGA, or field programmable gate array can then be custom-programmed by tablet computer vendors who have purchased these integrated circuits from the semiconductor device manufacturers.
[53] Sousou, Imad. "What's Next for MeeGo" (https://www.meego.com/community/blogs/imad/2011/whats-next-meego). meego.com. . Retrieved 28 September 2011.
[54] "Apple tablet rumors redux: 10.7-inch display, iPhone OS underneath" (http://www.engadget.com/2009/09/29/apple-tablet-rumors-redux-10-7-inch-display-iphone-os-undernea/). Engadget. . Retrieved August 6, 2010.
[55] Tifanny Boggs (2007) "Axiotron and OWC Unveil the ModBook" (http://www.tabletpcreview.com/default.asp?newsID=695)
[56] http://www.thinkyhead.com/tabletmagic
[57] BlackBerry PlayBook preview (http://www.youtube.com/watch?v=eAaez_4m9mQ)
[58] "9 Upcoming Tablet Alternatives to the Apple iPad" (http://mashable.com/2010/01/27/9-upcoming-tablet-alternatives-to-the-apple-ipad/). Mashable. . Retrieved August 7, 2010.
[59] "Don't bank on KMart's $150 Augen tablet getting Android Market access" (http://www.liliputing.com/2010/08/dont-bank-on-kmarts-150-augen-tablet-getting-android-market-access.html). liliputing. . Retrieved August 7, 2010.
[60] Toshiba debuts ultraslim Android laptop (http://news.cnet.com/8301-13924_3-20008301-64.html), 21 June 2010.
[61] "Forget all these Android tablets, let me at that Chrome OS" (http://www.crunchgear.com/2010/07/20/forget-all-these-android-tablets-let-me-at-that-chrome-os/). CrunchGear. . Retrieved August 7, 2010.
[62] "Google Chrome OS Tablet Brings Ties With Verizon" (http://www.informationweek.com/news/services/data/showArticle.jhtml?articleID=226700487)
[63] Motorola Android tablet in 2011 (http://www.marketwatch.com/video/asset/digits-motorola-plans-tablet-device-2010-09-16/7CC13B36-0A8B-42E0-AD1A-72FF9BF04348)

[64] Lenovo is waiting for Honeycomb (http://www.slashgear.com/tablets-a-prescription-for-confusion-24109978/) accessdate=2010-10-24
[65] The successor to *Gingerbread*, Android project *Honeycomb* is targeted for tablet computers. – Daniel Lyons (Oct. 11, 2010), *Newsweek* p. 49

- Google demonstrated Android Honeycomb Tablet 12/7/2010 (http://www.pcmag.com/article2/0,2817,2373943,00.asp)
- Andy Rubin's demo of Motorola Honeycomb tablet (http://www.marketwatch.com/video/asset/digits-best-buys-tv-bust-2010-12-14/FD6C213F-D320-4212-BE7C-FECD5B8FEA33?dist=afterbell#!88F98ADB-3F87-49DF-AD08-385D66B0DDE8)

[66] HP Webcast announcing the end of Touchpad, webOS devices (http://www.hp.com/investor/2011q3webcast) accessdate=2011-08-18
[67] HP announces that webOS and ENYO, its application development platform, are being contributed to the open-source community. (http://www.hp.com/hpinfo/newsroom/press/2011/111209xa.html) accessdate=2011-12-09
[68] One Laptop Gets $5.6M Grant From Marvell to Develop Next Generation Tablet Computer | Xconomy (http://www.xconomy.com/boston/2010/10/04/one-laptop-gets-5-6m-grant-from-marvell-to-develop-next-generation-tablet-computer/)
[69] This "vicious cycle" (slow hardware development masking slow hardware, causing slow response, causing slow software development, causing sluggish performance at an unrealistic price, causing sluggish sales) serves only to impede further software investment. "HP reboots to confront Tablet Effect" *Barron's*, August 20th, 2011
[70] Brown, Peter iPad is iBad for freedom (http://www.fsf.org/news/ibad_launch), Free Software Foundation, 2010
[71] Cherry, Steven The iPad Is Not a Computer (http://spectrum.ieee.org/consumer-electronics/portable-devices/the-ipad-is-not-a-computer), IEEE Spectrum, 2010
[72] Conlon, Tom The iPad's Closed System: Sometimes I Hate Being Right (http://www.popsci.com/gadgets/article/2010-01/ipadâs-closed-system-sometimes-i-hate-being-right), Popular Science, 2010
[73] Convertibles: The new laptop bling? – CNET News.com (http://news.com.com/Convertibles+The+new+laptop+bling/2100-1044_3-5900655.html)
[74] *PC World* (Nov 18, 2010 12:25 pm)"Why Tablet Computing Hasn't Been Big Business" (http://www.pcworld.com/businesscenter/article/211066/why_tablet_computing_hasnt_been_big_business.html)
[75] XO-3 concept design is here! | One Laptop per Child (http://blog.laptop.org/2009/12/24/xo-3-concept/)
[76] India unveils prototype for $35 touch-screen computer (http://www.bbc.co.uk/news/world-south-asia-10740817) BBC World news-South Asia Retrieved July 25, 2010
[77] India's ($)35 PC is the future of computing (http://www.pcworld.com/businesscenter/article/201769/indias_35_pc_is_the_future_of_computing.html?tk=hp_pop) PCWorld.com
[78] Bill of materials, *Wired* (http://www.wired.com/gadgetlab/2010/07/india-35-tablet/)
[79] "India launches "world's cheapest" tablet Aakash" (http://in.reuters.com/article/2011/10/05/idINIndia-59716920111005). *Reuters*. October 5, 2011..
[80] http://www.csmonitor.com/From-the-news-wires/2010/0723/35-computer-introduced-in-India $100 OLPC tablet computer
[81] Adam Shah (July 31, 2010), IDC, "Negroponte offers OLPC technology for $35 tablet" (http://www.goodgearguide.com.au/article/355270/negroponte_offers_olpc_technology_35_tablet/)
[82] OLPC X03 grant accessdate=2010-10-04 (http://www.xconomy.com/boston/2010/10/04/one-laptop-gets-5-6m-grant-from-marvell-to-develop-next-generation-tablet-computer/)

External links

- Tablet or Ultrabook? (http://www.talkingabouttech.com/?p=5)
- What makes a tablet a tablet? (FAQ) (http://news.cnet.com/8301-31021_3-20006077-260.html?tag=newsLeadStoriesArea.1) CNET.com May 28, 2010

Stylus_(computing)

In computing, a **stylus** (or **stylus pen**) is a small pen-shaped instrument that is used to input commands to a computer screen, mobile device or graphics tablet. With touchscreen devices, a user places a stylus on the surface of the screen to draw or make selections by tapping the stylus on the screen.[1]

A smartphone being operated with a stylus.

Pen-like input devices which are larger than a stylus, and offer increased functionality such as programmable buttons, pressure sensitivity and electronic erasers, are often known as digital pens.[1]

several styli; (L to R) PalmPilot Professional, Fossil Wrist PDA, Nokia 770, Audiovox XV6600, HP Jornada 520, Sharp Zaurus 5500, Fujitsu Lifebook P-1032

The stylus is the primary input device for personal digital assistants.[1] It is also used on the Nintendo DS and Nintendo 3DS game consoles.[2] Some smartphones, such as Windows Mobile phones, require a stylus for accurate input.[3] However, devices featuring multi-touch finger-input are becoming more popular than stylus-driven devices in the smartphone market;[4] capacitive styli, different from standard styli, can be used for these finger-touch devices (iPhone, etc.).

Graphics tablets use styli containing circuitry (powered by battery or operating passively by change of inductance), to allow multi-function buttons on the barrel of the pen or stylus to transmit user actions to the tablet. Some (probably most) tablets detect varying degrees of pressure sensitivity, e.g. for use in a drawing program to vary line thickness or color density.

The first use of a stylus pen in a computing device was the *Styalator*, demonstrated by Tom Dimond in 1957.[5]

See also

- Light pen
- Pen computing

References

[1] Shelly, Gary B.; Misty E. Vermaat (2009). *Discovering Computers: Fundamentals* (http://books.google.com.au/books?id=RT2nDtioi3sC&lpg=PT218&ots=XcwPsIjdBw&dq="a stylus is a"&pg=PT218#v=onepage&q=stylus&f=false). Cengage Learning. ISBN 9780495806387. . Retrieved 3 November 2009.

[2] "Giz Explains: The Magic Behind Touchscreens" (http://gizmodo.com/5036516/giz-explains-the-magic-behind-touchscreens). *Gizmodo*. 13 August 2008. . Retrieved 3 November 2009.

[3] Charles Arthur (20 October 2009). "Windows Mobile: where's the love? And where's the sales figure?" (http://www.guardian.co.uk/technology/blog/2009/oct/20/windows-mobile-reviews-negative). *The Guardian*. .

[4] "The Age of Touch Computing: A Complete Guide" (http://www.pcmag.com/article2/0,2817,2336835,00.asp). *PC World*. 15 December 2008. .

[5] Dimond, Tom (1957-12-01), *Devices for reading handwritten characters* (http://rwservices.no-ip.info:81/pens/biblio70.html#Dimond57), Proceedings of Eastern Joint Computer Conference, pp. 232–237, , retrieved 2008-08-23

Internationale_Funkausstellung_Berlin

Philips' stand on IFA 2009.

The **IFA** or **Internationale Funkausstellung Berlin** (International radio exhibition Berlin, aka 'Berlin Radio Show') is one of the oldest industrial exhibitions in Germany. Between 1924 and 1939 it was an annual event, but as from 1950 it was organized on a two yearly basis until 2005. Since then it has become an annual event again, held in September. Today it is one of world's leading trade shows for consumer electronics and home appliances.

It offers the opportunity to exhibitors to present their latest products and developments to the general public. As a result of daily reporting in almost all the German media, the radio exhibition achieves a large spreading of the information – and advertising messages and has also become international. In the course of its history, a large number of world innovations were first seen at the exhibition.[1] The next IFA will be held from August 31, to September 5, 2012 in Berlin.

History

Goebbels with a VE 301 W

German physicist and inventor Manfred von Ardenne gave a public demonstration of a television system using a cathode ray tube for both transmission (using flying-spot image scans, not a camera) and reception, at the 1931 show.[2]

In 1933 The Volksempfänger (VE 301 W),[3] a Nazi-sponsored radio receiver design, was introduced. Ordered by Dr. Joseph Goebbels, designed by Otto Griessing, sold by Gustav Seibt, it was presented at the tenth Berliner Funkausstellung on 18 August 1933[4] its price fixed at 76 Reichsmark (RM). 100,000 units were sold during the exhibition. The importance of radio as a mass medium for the dissemination of propaganda and entertainment throughout Germany had long been clear to Goebbels.

In 1938 the DKE 38 (*Deutscher Kleinempfänger 38*, i.e. German miniature receiver 1938) followed, the price fixed at 35 RM.

AEG, founded in 1883 by Emil Rathenau, showed the first practical audio tape recorder, the Magnetophon K1, at the August 1935 show.[5] [6]

In 1939 the exhibition was called *Grosse Deutsche Funk- und Fernseh-Ausstellung* (Great German Radio and Television Exhibition). The *Einheits-Fernseh-Empfänger E1*, a TV set designed to be affordable for everybody, was introduced. The physical display size was 7.68" × 8.86". Plans for large-scale manufacture were thwarted by the outbreak of WWII. Color TV was also introduced (a prototype), based on an invention by Werner Flechsig (cf. shadow mask).

Multinational Dutch electronics corporation Philips introduced the compact audio cassette medium for audio storage at the 1962 show.[7] [8] [9]

Main entrance and opening of exhibition in 1924

Funkkabine of the Z.R. 3.

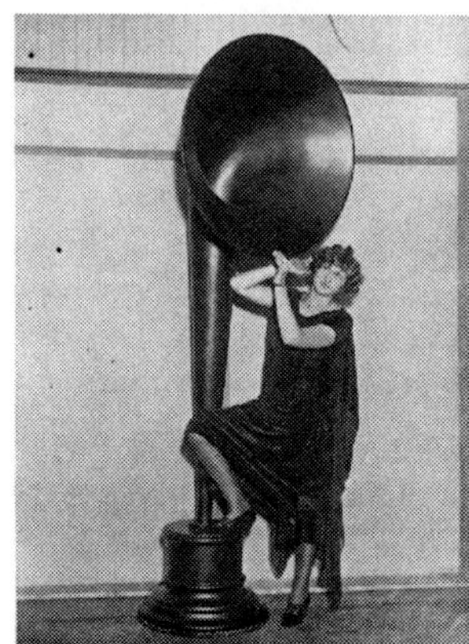

Presentation of a big speaker in 1929

Facts

- In 2003, IFA was the world's largest Consumer Electronics trade fair, attended by 273,800 visitors. Over 1,000 exhibitors attended the fair.
- 2.4 billion euro of placed orders.
- Attended by over 6,800 accredited journalists from 78 countries.[10]

Notes

[1] *Gramophone*: 206. November 1989. http://www.gramophone.net/Issue/Page/November%201989/206/766604/. Retrieved 4 August 2010.

[2] Albert Abramson, *Zworykin: Pioneer of Television*, University of Illinois Press, 1995, p. 111.

[3] VE=Volksempfänger; 301 = 30 January; W = Wechselstrom (=alternating current))

[4] Goebbel's speech (http://www.calvin.edu/academic/cas/gpa/goeb56.htm)

[5] History Department at the University of San Diego. "Magnetic Recording History Pictures" (http://history.sandiego.edu/gen/recording/tape.html). .

[6] "1935 AEG Magnetophon Tape Recorder" (http://mixonline.com/TECnology-Hall-of-Fame/aeg-magnetophone-recorder-090106/). mixonline.com. September 1, 2006. . Retrieved June 18, 2010.

[7] David Morton, *Sound recording: the life story of a technology*. Greenwood Publishing Group, 2004, p.161.

[8] John Shepherd, *Continuum encyclopedia of popular music of the world*. Continuum International Publishing Group, 2003, p.506

[9] "Cassette Rampage Forecast". *Billboard magazine* (Nielsen Business Media, Inc.) **79** (44): 1,72. November 4, 1967. ISSN 0006-2510.

[10] http://www1.messe-berlin.de/vip8_1/website/MesseBerlin/htdocs/www.ifa-berlin.de/fset_content_e.html?url=http://www1.messe-berlin.de/vip8_1/website/MesseBerlin/htdocs/www.ifa-berlin.de/en/Messeinfos/Profil/Kurzbeschreibung/index.html

External links

- IFA official website (http://www1.messe-berlin.de/vip8_1/website/Internet/Internet/www.ifa-berlin/englisch/index.html)
- Highlights of past exhibitions from 1926 to 2005 (http://berlin.barwick.de/events/ifa-internationale-funkausstellung.html#highlights)
- IFA Chronicle – multimedia based highlights (http://www1.messe-berlin.de/vip8_1/website/Internet/Internet/www.ifa-berlin/englisch/ABOUT-IFA/Profile/IFA-Chronicle/index.jsp)

AT&T

Type	Public
Traded as	NYSE: T [1] NYSE: ATT [2] Dow Jones Component S&P 500 Component
Industry	Telecommunications
Founded	October 5, 1983[3]
Headquarters	Whitacre Tower, Downtown, Dallas, Texas, United States
Key people	Randall Stephenson (Chairman, President & CEO)[4]
Services	Fixed line and mobile telephony, broadband and fixed-line internet services, digital television
Revenue	▲ US$ 124.280 billion (2010)[5]
Operating income	▼ US$ 19.573 billion (2010)[5]
Net income	▲ US$ 19.864 billion (2010)[5]
Total assets	▲ US$ 277.653 billion (2011)[5]
Total equity	▲ US$ 111.950 billion (2010)[5]
Employees	294,600 (2010)[5]
Parent	AT&T Corp. (1983)
Subsidiaries	AT&T Corp. AT&T Mobility BellSouth Southwestern Bell AT&T Teleholdings
Website	ATT.com [6]

The horizontal version on the AT&T logo, more commonly seen than the vertical version.

AT&T Inc. (sometimes stylized as **at&t**; NYSE: T [1], for "telephone") is an American multinational telecommunications corporation headquartered in Whitacre Tower, Dallas, Texas, United States. It is the second largest provider of mobile telephony and fixed telephony in the United States, and is also a provider of broadband and subscription television services. As of 2010, AT&T is the 7th largest company in the United States by total revenue, as well as the 4th largest non-oil company in the US (behind Walmart, General Electric, and Bank of America). It is the 3rd largest company in Texas by total revenue (behind ExxonMobil and ConocoPhillips) and the largest non-oil company in Texas. It is also the largest company headquartered in Dallas.[7] In 2011, Forbes listed AT&T as the 14th largest company in the world by market value[8] and the 9th largest non-oil company in the world by market value.[9] It is the 20th largest mobile telecom operator in the world with over 100.7 million mobile customers.[10]

Whitacre Tower: AT&T's corporate headquarters in Dallas, Texas

The company began its existence as **Southwestern Bell Corporation**, one of seven Regional Bell Operating Companies created in 1983 as part of the break-up of the original AT&T due to the *United States v. AT&T* antitrust lawsuit. It changed its name to **SBC Communications Inc.** in 1995. In 2005, it purchased its former parent company, AT&T Corporation, and took on its branding, with the merged entity naming itself *AT&T Inc.* and using the iconic AT&T logo and stock-trading symbol.

The current AT&T reconstitutes much of the former Bell System and includes ten of the original 22 Bell Operating Companies, along with one it partially owned (Southern New England Telephone), and the original long distance division.[11] The company is headquartered in downtown Dallas, Texas.[12]

History

1984–1995: Southwestern Bell Corporation

American Telephone and Telegraph Company officially transferred full ownership of Southwestern Bell Telephone Company to Southwestern Bell Corporation on January 1, 1984. It had three other subsidiaries: Southwestern Bell Publications, Inc., a directory publisher; Southwestern Bell Mobile Systems, Inc., in the business of mobile telephone service; and Southwestern Bell Telecommunications, Inc., focusing on marketing phone equipment to business customers. The holding companies' new president was Zane Edison Barnes.

Southwestern Bell Corporation logo, 1992-1995

In 1987, SBC bought Metromedia Inc.'s cellular and paging business. This in turn boosted the company to third largest cellular-communications company in the United States; behind McCaw Cellular and Pacific Telesis. In January 1990, Edward Whitacre took over as president of Southwestern Bell. The Headquarters was moved from St. Louis to San Antonio, Texas in February 1993. It acquired 2 cable companies in Maryland and Virginia from Hauser Communications for $650 million, becoming the first regional Bell telephone company to acquire a cable company outside of its service area. In 1994, they called off a $1.6 billion acquisition attempt for 40% of Cox Cable due to FCC rules on cable companies. SBC would later start selling its current cable company interests.

1995–2000: Changes in the company

In 1995 Southwestern Bell Corp. became SBC Communications. They then combined Southwestern Bell Telecom division (which made telephone equipment) into the company, due to new FCC rules.

SBC Communications logo, 2001-2005

In 1996, SBC announced it would acquire Pacific Telesis Group, a Regional Bell Operating Company (RBOC) in California and Nevada. 1997 brought rumors of a proposed merger between AT&T Corporation (the USA's largest long distance provider) and SBC (the USA's largest local provider). The FCC disapproved of the merger, and it came to end. Later in 1997, SBC sold its last two cable companies, exiting the cable telecom field.

In January 1998, SBC announced it would take over Southern New England Telecommunications Corp. (SNET) for $4.4 billion in stock (the FCC would approve in October 1998). SBC also won a court judgment that would make it easier for RBOCs to enter the long distance phone service, but it was being challenged by AT&T and the FCC. In May 1998, Ameritech and SBC announced a $62 billion dollar merger, in which SBC would take over Ameritech. After making several organizational changes (such as the sale of Ameritech Wireless to GTE) to satisfy state and federal regulators, the two merged on October 8, 1999. The FCC later fined SBC Communications $6 million for failure to comply with agreements made in order to secure approval of the merger. SBC became the largest RBOC until the Bell Atlantic and GTE merger. 1998 revenues were $46 billion, placing SBC among the top 15 companies in the Fortune 500.

In January 1999, SBC announced it would purchase Comcast Cellular, for $1.7 billion, plus $1.3 billion of debt. During 1999 SBC continued to prepare to be allowed to provide long distance phone service. February SBC acquired up to ten percent of Williams Companies' telecommunications division for about $500 million, who was building a fiber optic network across the country and would carry SBC's future service. On November 1, 1999, SBC became a part of the Dow Jones Industrial Average.

2000–2005: One national brand, and acquisition of AT&T Corporation

In 2002, SBC ended marketing its operating companies under different names, and simply opted to give its companies different doing business as names based on the state (a practice already in use by Ameritech since 1993), and it gave the holding companies it had purchased d/b/a names based on their general region.

The final AT&T logo prior to its merger with SBC in late 2005.

SBC-AT&T legacy transition logo, used 2005–2006

On January 31, 2005, SBC announced that it would purchase AT&T Corporation for more than US$16 billion. The announcement came almost eight years after SBC and AT&T (also known as American Telephone & Telegraph Corp.) called off their first merger talks and nearly a year after initial merger talks between AT&T Corp. and BellSouth fell apart. AT&T stockholders meeting in Denver, approved the merger on June 30, 2005. The U.S. Department of Justice cleared the merger on October 27, 2005, and the Federal Communications Commission approved it on October 31, 2005.

The merger was finalized on November 18, 2005.[13] Upon the completion of the merger, SBC Communications adopted the AT&T branding, and changed its corporate name to **AT&T Inc.** to differentiate the company from the former AT&T Corporation. On December 1, 2005, the merged company's New York Stock Exchange ticker symbol was changed from "SBC" to the traditional "T" used by AT&T.

The new AT&T updated the former AT&T's graphic logo (a new "marble" designed by Interbrand took over the "Death Star"); however the existing AT&T sound trademark (voiced by Pat Fleet) continues to be used.

2006: BellSouth acquisition

The BellSouth logo.

On Friday December 29, 2006, the Federal Communications Commission (FCC) approved the new AT&T's acquisition of a regional Bell Operating Company, BellSouth, valued at approximately $86 billion (or 1.325 shares of AT&T for each share of BellSouth at the close of trading December 29, 2006).[14] The new combined company retained the name AT&T.[15] The deal consolidated ownership of both Cingular Wireless and Yellowpages.com, once joint ventures between BellSouth and AT&T. All services, including wireless, became offered under the AT&T name.[16]

2007–2008 restructuring

Transition to new media

The AT&T Switching Center in downtown Los Angeles.

In June 2007, AT&T's new chairman and CEO, Randall Stephenson, discussed how wireless services are the core of "The New AT&T".[17] With declining sales of traditional home phone lines, AT&T plans to roll out various new media such as Video Share, U-verse, and to extend its reach in high speed Internet into rural areas across the country. AT&T announced on June 29, 2007, however, that it was acquiring Dobson Communications. It was then reported on October 2, 2007 that AT&T would purchase Interwise for $121 million, which it completed on November 2, 2007. On October 9, 2007, AT&T purchased 12 MHz of spectrum in the prime 700 MHz spectrum band from privately held Aloha Partners for nearly $2.5 billion; the deal was approved by the FCC on February 4, 2008. On December 4, 2007 AT&T announced plans to acquire Edge Wireless, a regional GSM carrier in the Pacific Northwest.[18] The Edge Wireless acquisition was completed in April 2008.[19]

Corporate headquarters move

On June 27, 2008, AT&T announced that it would move its corporate headquarters from 175 East Houston Street in San Antonio to One AT&T Plaza in Downtown Dallas.[12] [20] The company said that it moved to gain better access to its customers and operations throughout the world, and to the key technology partners, suppliers, innovation and human resources needed as it continues to grow, domestically and internationally[21]

AT&T Inc. previously relocated its corporate headquarters to San Antonio from St. Louis in 1992, when it was then named Southwestern Bell Corporation. The company's Telecom Operations group, which serves residential and regional business customers in 22 U.S. states, remains in San Antonio.

Atlanta continues to be the headquarters for AT&T Mobility, with significant offices in Redmond, Washington, the former home of AT&T Wireless. Bedminster, New Jersey is the headquarters for the company's Global Business Services group and AT&T Labs. St. Louis continues as home to the company's Directory operations, AT&T Advertising Solutions.[22]

Job cuts

On December 4, 2008, AT&T announced they would be cutting 12,000 jobs due to "economic pressures, a changing business mix and a more streamlined organizational structure".[23]

Post-consolidation wireless acquisitions

Cellular One acquisition

On June 29, 2007 AT&T announced that they had reached an agreement to purchase Dobson Cellular, which provided services in the US under the name Cellular One in primarily rural areas. The closing price was $2.8B USD, or $13 per share. AT&T also agreed to assume the outstanding debt of $2.3B USD. The sale completed on November 15, 2007, with market transition beginning December 9, 2007.[24]

Centennial acquisition

On November 11, 2008, AT&T announced a $944 million buyout of Centennial Communications Corp. The acquisition is subject to regulatory approval, the approval of Centennial's stockholders and other customary closing conditions. Welsh, Carson, Anderson & Stowe, Centennial's largest stockholder, has agreed to vote in support of this transaction. In an attempt to quell regulators, on May 9, 2009 AT&T entered an agreement with Verizon Wireless to sell off certain existing Centennial service areas in the states of Louisiana and Mississippi for $240 million pending the successful merger of AT&T and Centennial.[25]

Wayport acquisition

On December 12, 2008, AT&T acquired Wayport, Inc., a major provider of Internet hotspots in the United States. With the acquisition, AT&T's public Wi-Fi deployment climbed to 20,000 hotspots in the United States, the most of any U.S. provider.[26]

Qualcomm spectrum

On December 20, 2011, AT&T and Qualcomm announced that AT&T would buy $1.93 billion worth of spectrum from Qualcomm. Formerly used for FLO TV, this spectrum will be used to expand AT&T's 4G wireless services. AT&T already had spectrum for the purpose close to what it is buying.[27]

Attempted acquisition of T-Mobile USA

On March 20, 2011, AT&T announced its intention to buy T-Mobile USA for $39 billion from Deutsche Telekom. The deal comes with 33.7 million subscribers, making AT&T the largest mobile phone company in the United States.[28] [29] If the deal goes through AT&T would have a 43% market share of mobile phones in the U.S. making AT&T significantly larger than any of its competitors. Regulators question the effects such a deal will have on both competitors and consumers.[28] AT&T CEO Randall Stephenson however stated that the merger would increase network quality and would lead to large savings for the company. AT&T stated it may have to sell some assets to gain approval from regulators, but claims to have done their "homework" on regulations.[30]

Reaction to the announced merger has generated both support as well as opposition among various groups and communities.

The merger has garnered support from a wide number of civil rights, environmental, and business organizations. These include the NAACP, League of United Latin American Citizens, Gay & Lesbian Alliance Against Defamation (GLAAD), League of United Latin American Citizens (LULAC), and the Sierra Club.[31] Labor organizations such as the AFL-CIO, Teamsters, and the Communications Workers of America also voiced support for the merger. These organizations point to AT&T's commitment to labor, social, and environmental standards. Many of these organizations have also cited how the merger is likely to accelerate 4G wireless deployment, thus helping underserved communities such as rural areas and disadvantaged urban communities.[31] According to the NAACP, the merger will "advance increased access to affordable and sustainable wireless broadband services and in turn stimulate job creation and civic engagement throughout our country."[31]

As of August 2 the governors of 26 states have written letters supporting the merger.[32] On July 27 the attorneys general of Utah, Alabama, Arkansas, Georgia, Kentucky, Michigan, Mississippi, North Dakota, South Dakota, West Virginia, and Wyoming sent a joint letter of support to the FCC.[32] As of August 2011 state regulatory agencies in Arizona and Louisiana have approved the acquisition.

A diverse group of industry and public-interest organizations are opposed to AT&T's merger with T-Mobile. Consumer groups including Public Knowledge, Consumers Union, Free Press and the Media Access Project are publicly opposed to AT&T merger. These groups have influence with Democrats at the Federal Communications Commission and in Congress. These organizations fear that the merger will raise prices and stifle innovation by consolidating so much of the wireless industry in one company. Free Press and Public Knowledge have started letter-writing campaigns against the deal.[33]

Internet companies are generally skeptical of the merger because it leaves them with fewer counter-parties to negotiate with for getting their content and applications to customers. The AT&T merger might leave them dependent on just two, AT&T and Verizon. The Computer & Communication Industry Association (CCIA), which counts Google, Microsoft, Yahoo and eBay among its members, is opposed to the merger. "A deal like this, if not blocked on antitrust grounds, is of deep concern to all the innovative businesses that build everything from apps to handsets. It would be hypocritical for our nation to talk about unleashing innovation on one hand and then stand by as threats to innovation like this are proposed," said Ed Black, head of CCIA.[33]

On April 21, 2011, AT&T defended its proposed acquisition of T-Mobile USA before a U.S. Senate committee, saying the combined company will deliver high-speed wireless services to 97 percent of Americans and provide consumer benefits such as fewer dropped calls.[34]

If AT&T's acquisition of T-Mobile USA is rejected by federal regulators, AT&T would need to pay $6 billion, including $3 billion in cash, to T-Mobile USA's parent company Deutsche Telekom.[35]

On August 31, 2011, the Department of Justice officially filed a lawsuit in the United States District Court for the District of Columbia seeking to block the acquisition. [36] [37]

On November 30, 2011, the FCC allowed AT&T to withdraw their merger, saving both carriers from divulging documentation about internal operations. The FCC cited job loss and higher consumer prices as reasons to deny the merger.[38]

On December 19, 2011, AT&T announced that it would permanently end its merger bid after a "thorough review of its options". The announcement included an assertion that the failure of the acquisition would increase costs to consumers and harm innovation in the wireless market. As per the original acquisition agreement, T-Mobile will receive $3 billion in cash as well as access to $1 billion worth of AT&T-held wireless spectrum.[39]

Political contributions and lobbying

According to the Center for Responsive Politics, AT&T is the second largest donor to United States political campaigns,[40] and the top American corporate donor,[41] having contributed more than US$47.7 million since 1990, 56% and 44% of which went to Republican and Democratic recipients, respectively.[42] Also, during the period of 1998 to 2010, the company expended US$130 million on lobbying in the United States.[41] A key political issue for AT&T has been the question of which businesses win the right to profit by providing broadband internet access in the United States.[43]

In 2005, AT&T was among 53 entities that contributed the maximum of $250,000 to the second inauguration of President George W. Bush.[44] [45] [46]

Bell Operating Companies

Of the twenty-two Bell Operating Companies which AT&T Corp. owned prior to the 1984 agreement to divest, eleven (BellSouth Telecommunications combines two former BOCs) have become a part of the new AT&T Inc. with the completion of their acquisition of BellSouth Corporation on December 29, 2006:[47]

AT&T payphone signage.

- BellSouth Telecommunications (formerly known as Southern Bell; includes former South Central Bell)
- Illinois Bell
- Indiana Bell
- Michigan Bell
- Nevada Bell (formerly known as Bell Telephone Company of Nevada)
- Ohio Bell
- Pacific Bell (formerly Pacific Telephone & Telegraph)
- Southwestern Bell
- Wisconsin Bell (formerly Wisconsin Telephone)
- Southern New England Telephone – Now wholly owned; the original AT&T held 16.8% interest prior to 1984.

Former operating companies

The following companies have gone to defunct status under SBC/AT&T ownership:

- Southwestern Bell Texas – a separate operating company created by SBC, absorbed operations of original SWBT on December 30, 2001 and became **Southwestern Bell Telephone, L.P.**; eventually merged into **SWBT Inc.** in 2007 which became the current Southwestern Bell
- Woodbury Telephone – merged into Southern New England Telephone on June 1, 2007.

Corporate structure

AT&T Inc. has retained the holding companies it has acquired over the years resulting in the following corporate structure:

AT&T office in San Antonio, Texas with new logo and orange highlight from the former Cingular

- AT&T Inc., publicly traded holding company
  - Southwestern Bell Telephone Company d/b/a AT&T Arkansas, AT&T Kansas, AT&T Missouri, AT&T Oklahoma, AT&T Southwest, AT&T Texas
  - AT&T Teleholdings, Inc. d/b/a AT&T East, AT&T Midwest, AT&T West; formerly Ameritech, acquired in 1999; absorbed Pacific Telesis and SNET Corp. under AT&T ownership
    - Illinois Bell Telephone Company d/b/a AT&T Illinois
    - Indiana Bell Telephone Company d/b/a AT&T Indiana
    - Michigan Bell Telephone Company d/b/a AT&T Michigan
    - The Ohio Bell Telephone Company d/b/a AT&T Ohio
    - Pacific Bell Telephone Company d/b/a AT&T California
      - Nevada Bell Telephone Company d/b/a AT&T Nevada
    - The Southern New England Telephone Company d/b/a AT&T Connecticut (includes former Woodbury Telephone)
    - Wisconsin Bell, Inc. d/b/a AT&T Wisconsin
  - AT&T Corp., acquired 2005
    - AT&T Alascom
  - BellSouth Corporation d/b/a AT&T South, acquired 2006
    - BellSouth Telecommunications, LLC d/b/a AT&T Alabama, AT&T Florida, AT&T Georgia, AT&T Louisiana, AT&T Kentucky, AT&T Mississippi, AT&T North Carolina, AT&T South Carolina, AT&T Southeast, AT&T Tennessee
  - AT&T Mobility

Corporate governance

AT&T's current board of directors:[48]

Stephenson at the 2008 World Economic Forum

- Randall L. Stephenson – Chairman and Chief Executive Officer
- James A. Henderson
- Gilbert F. Amelio
- Reuben V. Anderson
- James H. Blanchard
- Jaime Chico Pardo
- James P. Kelly
- Jon C. Madonna
- Lynn M. Martin
- John B. McCoy
- Joyce M. Roché
- Matthew K. Rose
- Laura D'Andrea Tyson

Criticism and controversies

Wireless service

AT&T has received criticisms for its wireless services. In December 2010, *Consumer Reports* named AT&T as the worst wireless provider in the country.[49] In 2011, AT&T has been rated the worst wireless provider for the second year in a row.[50]

Censorship

In August 2009, the band Pearl Jam performed in Chicago at Lollapalooza which was being web-broadcast by AT&T. The band, while playing the song "Daughter", started playing a version of Pink Floyd's "Another Brick in the Wall" but with altered lyrics critical of president George Bush. These lyrics included "George Bush, leave this world alone!" and, "George Bush, find yourself another home!". Listeners to AT&T's web broadcast heard only the first line because the rest was censored,[51] although AT&T spokesman Michael Coe said that the silencing was "a mistake."[52]

In September 2007, AT&T changed[53] their legal policy to state that "AT&T may immediately terminate or suspend all or a portion of your Service, any Member ID, electronic mail address, IP address, Universal Resource Locator or domain name used by you, without notice for conduct that AT&T believes"..."(c) tends to damage the name or reputation of AT&T, or its parents, affiliates and subsidiaries."[54] By October 10, 2007 AT&T had altered the terms and conditions for its Internet service to explicitly support freedom of expression by its subscribers, after an outcry claiming the company had given itself the right to censor its subscribers' transmissions.[55]

Section 5.1 of AT&T's new terms of service now reads "AT&T respects freedom of expression and believes it is a foundation of our free society to express differing points of view. AT&T will not terminate, disconnect or suspend service because of the views you or we express on public policy matters, political issues or political campaigns."[56]

On July 26, 2009, AT&T customers were unable to access certain sections of the image board 4chan, specifically /b/ (the "random" board) and /r9k/ (the "ROBOT 9000" board, a spin-off of the random board).[57] However, by the morning of Monday, July 27, the block had been lifted and access to the affected boards was restored. AT&T's official reason for the block was that a distributed denial of service attack had originated from the img.4chan.org server, and access was blocked to stop the attack.[58] Major news outlets have reported that the issue may be related to DDoSing of 4chan, and that the suspicions of 4chan users fell on Kimmo Alm, the owner of AnonTalk.com (later AnonTalk.se) at that time for doing this.[59] Alm has been reported in the past to have DDoSed 4chan.[60]

Privacy controversy

Further information: NSA call database, Mark Klein, NSA warrantless surveillance controversy, Hepting v. AT&T

In 2006, the Electronic Frontier Foundation lodged a class action lawsuit, *Hepting v. AT&T*, which alleged that AT&T had allowed agents of the National Security Agency (NSA) to monitor phone and Internet communications of AT&T customers without warrants. If true, this would violate the Foreign Intelligence Surveillance Act of 1978 and the First and Fourth Amendments of the U.S. Constitution. AT&T has yet to confirm or deny that monitoring by the NSA is occurring. In April 2006, a retired former AT&T technician, Mark Klein, lodged an affidavit supporting this allegation.[62] [63] The Department of Justice has stated they will intervene in this lawsuit by means of State Secrets Privilege.[64] In July 2006, the United States District Court for the Northern District of California – in which the suit was filed – rejected a federal government motion to dismiss the case. The motion to dismiss, which invoked the State Secrets Privilege, had argued that any court review of the alleged partnership between the federal government and AT&T would harm national security. The case was immediately appealed to the Ninth Circuit. It was dismissed on June 3, 2009, citing retroactive legislation in the Foreign Intelligence Surveillance Act. [65] In May 2006, *USA Today* reported that all international and domestic calling records had been handed over to the National Security Agency by AT&T, Verizon, SBC, and BellSouth for the purpose of creating a massive calling database.[66] The portions of the *new* AT&T that had been part of SBC Communications before November 18, 2005 were not mentioned.

Exhibit A

Diagram of how alleged wiretapping worked. From EFF court filings[61]

On June 21, 2006, the *San Francisco Chronicle* reported that AT&T had rewritten rules on their privacy policy. The policy, which took effect June 23, 2006, says that "*AT&T – not customers – owns customers' confidential info and can use it 'to protect its legitimate business interests, safeguard others, or respond to legal process.'*"[67]

On August 22, 2007, National Intelligence Director Mike McConnell confirmed that AT&T was one of the telecommunications companies that assisted with the government's warrantless wire-tapping program on calls between foreign and domestic sources.[68]

On November 8, 2007, Mark Klein, a former AT&T technician, told Keith Olbermann of MSNBC that all Internet traffic passing over AT&T lines was copied into a locked room at the company's San Francisco office – to which only employees with National Security Agency clearance had access.[69]

AT&T keeps for five to seven years a record of who text messages whom and the date and time, but not the content of the messages.[70]

Intellectual property filtering

In January 2008, the company reported plans to begin filtering all Internet traffic which passes through its network for intellectual property violations.[71] Commentators in the media have speculated that if this plan is implemented, it would lead to a mass exodus of subscribers leaving AT&T,[72] although this is misleading as Internet traffic may go through the company's network anyway.[71] Internet freedom proponents used these developments as justification for government-mandated network neutrality.

Discrimination against local Public-access television channels

AT&T is accused by community media groups of discriminating against local Public, educational, and government access (PEG) cable TV channels:, by "imposing unfair restrictions that will severely restrict the audience".[73]

According to Barbara Popovic, Executive Director of the Chicago public-access service CAN-TV, the new AT&T U-verse system forces all Public-access television into a special menu system, denying normal functionality such as channel numbers, access to the standard program guide, and DVR recording.[73] The Ratepayer Advocates division of the California Public Utilities Commission reported: "Instead of putting the stations on individual channels, AT&T has bundled community stations into a generic channel that can only be navigated through a complex and lengthy process."[73]

Sue Buske (president of telecommunications consulting firm the Buske Group and a former head of the National Federation of Local Cable Programmers/Alliance for Community Media) argue that this is "an overall attack [...] on public access across the [United States], the place in the dial around cities and communities where people can make their own media in their own communities".[73]

Information security

In June 2010, a hacker group known as Goatse Security discovered a vulnerability within the AT&T that could allow anyone to uncover email addresses belonging to customers of AT&T 3G service for the Apple iPad.[74] These email addresses could be accessed without a protective password.[75] Using a script, Goatse Security collected thousands of email addresses from AT&T.[74] Goatse Security informed AT&T about the security flaw through a third party.[76] Goatse Security then disclosed around 114,000 of these emails to Gawker Media, which published an article about the security flaw and disclosure in *Valleywag*.[74] [76] Praetorian Security Group criticized the web application that Goatse Security exploited as "poorly designed".[74]

Naming rights and sponsorships

Buildings

- AT&T 220 Building – building in Indianapolis, Indiana
- AT&T Building – building in Detroit, Michigan
- AT&T Building – building in Indianapolis, Indiana
- AT&T Building – building in Kingman, Arizona
- AT&T Building – (aka "The Batman Building") in Nashville, Tennessee
- AT&T Building – building in Omaha, Nebraska
- AT&T Building Addition – building in Detroit, Michigan
- AT&T Center – building in Los Angeles
- AT&T Center – building in St. Louis, Missouri
- AT&T City Center – building in Birmingham, Alabama
- AT&T Corporate Center – building in Chicago, Illinois
- AT&T Huron Road Building – building in Cleveland, Ohio
- AT&T Lenox Park Campus – AT&T Mobility Headquarters in DeKalb County just outside Atlanta, Georgia
- AT&T Midtown Center – building in Atlanta, Georgia
- AT&T Switching Center – building in Los Angeles
- AT&T Switching Center – building in Oakland, California
- AT&T Switching Center – building in San Francisco

AT&T Midtown Center in Atlanta, Georgia

- AT&T Building – building in San Diego
- Whitacre Tower (One AT&T Plaza) – Corporate Headquarters, Dallas, Texas
- Sony Tower, (formerly the *AT&T Building*)
- AT&T Tower – building in Jacksonville, FL

Venues

- AT&T Bricktown Ballpark – Oklahoma City, Oklahoma (formerly *Southwestern Bell Bricktown Ballpark, SBC Bricktown Ballpark)*
- AT&T Center – San Antonio, Texas (formerly *SBC Center*)
- AT&T Field – Chattanooga, Tennessee (formerly *BellSouth Park*)
- AT&T Park – San Francisco (formerly *Pacific Bell Park, SBC Park*)
- AT&T Plaza – Chicago, Illinois (public space that hosts the Cloud Gate sculpture in Millennium Park)
- AT&T Plaza – Dallas, Texas (plaza in front of the American Airlines Center at Victory Park)
- AT&T Performing Arts Center – Dallas, Texas
- Jones AT&T Stadium – Lubbock, Texas (formerly *Clifford B. and Audrey Jones Stadium, Jones SBC Stadium*)
- TPC San Antonio – San Antonio, Texas (AT&T Oaks Course & AT&T Canyons Course)

AT&T Center in San Antonio, Texas

Sponsorships

- AT&T Champions Classic – Valencia, California
- AT&T Classic – Atlanta, Georgia (formerly *BellSouth Classic*)
- AT&T Cotton Bowl Classic (formerly *Mobil Cotton Bowl Classic, Southwestern Bell Cotton Bowl Classic, SBC Cotton Bowl Classic*) – played in Arlington, Texas, at Cowboys Stadium.
- AT&T National – Washington, D.C.
- AT&T Pebble Beach National Pro-Am
- AT&T Red River Rivalry – Dallas, Texas (formerly *Red River Shootout, SBC Red River Rivalry*)
- Major League Soccer and the United States Soccer Federation, including the U.S. men's and U.S. women's national teams and the Major League Soccer All-Star Game from 2009
- United States Olympic team[77]
- National Collegiate Athletic Association (Corporate Champion)[78]

AT&T sponsors the annual Red River Rivalry football game

Miscellaneous

- AT&T (SEPTA station) – Public Transportation Station in Philadelphia, PA

Global presence

AT&T offers services in many locations throughout the Asia Pacific; its regional headquarters is located in Hong Kong.[79]

Enterprise SIP Trunking Services

In 2008, Toshiba announced SIP interoperability with the AT&T IP Flexible Reach service, and this partnership would improve Toshiba's IP-PBX offerings.[80] In 2010, AT&T combined its virtual private network (VPN) service with its IP Flexible Reach.[81]

See also

- AT&T Corporation
- AT&T Connect
- AT&T Mobility
- att.net
- Bell System
- Bell System Divestiture
- Communications Assistance For Law Enforcement Act
- Hepting v. AT&T
- Lists of public utilities
- Modification of Final Judgment
- NSA warrantless surveillance
- PRX (telephony)
- Regional Bell Operating Company
- Toktumi
- Tying of the iPhone to AT&T

References

[1] http://www.nyse.com/about/listed/quickquote.html?ticker=t
[2] http://www.nyse.com/about/listed/quickquote.html?ticker=att
[3] "Sec 8-k" (http://yahoo.brand.edgar-online.com/EFX_dll/EDGARpro.dll?FetchFilingHTML1?SessionID=jaJ3juNdQ_X7dSV&ID=4390374) (Press release). AT&T. April 28, 2006. . Retrieved September 29, 2007.
[4] "Randall L. Stephenson, Chairman, Chief Executive Officer and President" (http://www.att.com/gen/investor-relations?pid=7824). . Retrieved August 14, 2011.
[5] "2010 Form 10-K, AT&T Inc" (http://www.sec.gov/Archives/edgar/data/732717/000073271711000014/ex13.htm). United States Securities and Exchange Commission. .
[6] http://www.att.com
[7] "Fortune 500 2010: States: Texas Companies - FORTUNE on CNNMoney.com" (http://money.cnn.com/magazines/fortune/fortune500/2010/states/TX.html). Money. May 3, 2010. . Retrieved August 14, 2010.
[8] "AT&T" (http://www.forbes.com/companies/att/). *Forbes*. . Retrieved June 6, 2011.
[9] "The World's Biggest Public Companies" (http://www.forbes.com/global2000/). *Forbes*. . Retrieved June 6, 2011.
[10] "AT&T Reports Record 2.8 Million Wireless Net Adds, Strong U-verse Sales, Continued Revenue Gains in the Fourth Quarterl Europe, Middle East, Africal AT&T" (http://www.corp.att.com/emea/insights/pr/eng/q4_270111.html). Corp.att.com. . Retrieved 2011-11-28.
[11] Kleinfield, Sonny (1981). *The biggest company on earth: a profile of AT&T*. New York: Holt, Rinehart, and Winston. ISBN 978-0-03-045326-7.
[12] Godinez, Victor and David McLemore. " AT&T moving headquarters to Dallas from San Antonio (http://www.dallasnews.com/sharedcontent/dws/bus/stories/DN-att_28bus.ART.State.Edition2.4d5475b.html)." *The Dallas Morning News*. Saturday June 28, 2008.

Retrieved on June 18, 2009.
[13] "New AT&T Launches" (http://www.att.com/gen/press-room?pid=4800&cdvn=news&newsarticleid=21906) (Press release). AT&T. November 18, 2005. . Retrieved September 29, 2007.
[14] Vorman, Julie (December 29, 2006). "AT&T closes $86 billion BellSouth deal" (http://www.reuters.com/article/ousiv/idUSWBT006361 20070102). Reuters. . Retrieved September 29, 2007.
[15] Bartash, Jeffry; Jonathan Burton (March 5, 2006). "AT&T to pay $67 billion for BellSouth" (http://www.marketwatch.com/News/Story/Story.aspx?guid={6E4D6E93-004F-4938-9692-B2704970428B}&siteid=mktw&dist=). Dow Jones Market Watch. . Retrieved September 29, 2007.
[16] "AT&T and BellSouth Join to Create a Premier Global Communications Company" (http://www.att.com/gen/press-room?pid=4800&cdvn=news&newsarticleid=22860) (Press release). AT&T. December 29, 2006. . Retrieved September 29, 2007.
[17] Johnson, K C (June 24, 2007). "AT&T's new chief dialed in" (http://www.chicagotribune.com/business/chi-sun_front_0624jun24,0,3337832.story?track=rss). *Chicago Tribune*. . Retrieved June 27, 2007.
[18] "AT&T Buys Edge Wireless" (http://www.phonescoop.com/news/item.php?n=2569). Phone Scoop. December 4, 2007. . Retrieved July 26, 2008.
[19] Antonio, San (April 18, 2008). "AT&T completes buy out of Edge Wireless" (http://www.bizjournals.com/sanantonio/stories/2008/04/14/daily37.html). American City Business Journals. . Retrieved July 26, 2008.
[20] " Corporate Inquiries (http://www.att.com/gen/press-room?pid=1916)." *AT&T*. Retrieved on March 25, 2009.
[21] Source: Dallas News (http://www.dallasnews.com/sharedcontent/dws/dn/latestnews/stories/062808dnbusattmove.4515fb49.html)
[22] AT&T – News Room (June 27, 2008). AT&T Corporate Headquarters to Move to Dallas (http://www.att.com/gen/press-room?pid=4800&cdvn=news&newsarticleid=25882). Press release. Retrieved on June 27, 2008.
[23] "Job cuts mount as year-end nears" (http://money.cnn.com/2008/12/04/news/companies/ATNT/index.htm?postversion=2008120409) (Press release). AT&T. December 4, 2008. .
[24] "AT&T Completes Acquisition of Dobson Communications to Enhance Wireless Coverage" (http://www.att.com/gen/press-room?pid=4800&cdvn=news&newsarticleid=24739) (Press release). Dallas, Texas: AT&T Inc.. November 15, 2007. . Retrieved May 11, 2009.
[25] "AT&T Agrees to Sell Certain Centennial Communications Corp. Assets to Verizon Wireless" (http://newsticker.welt.de/?module=smarthouse&id=885963=) (Webpage). www.newsticker.welt.de. April 9, 2009. . Retrieved May 9, 2009.
[26] "AT&T Advanced Wi-Fi Strategy" (http://www.wayport.com/NewsReleases.aspx?id=2030) (Press release). Dallas, Texas: AT&T Inc.. December 12, 2008. . Retrieved December 22, 2008.
[27] "AT&T buys $2 billion worth of 4G spectrum from Qualcomm" (http://www.news-record.com/content/2010/12/20/article/att_buys_2_billion_worth_of_4g_spectrum_from_qualcomm). *News & Record*. Associated Press. December 20, 2010. . Retrieved December 20, 2010.
[28] "AT&T to Buy T-Mobile USA for $39 billion" (http://dealbook.nytimes.com/2011/03/20/att-to-buy-t-mobile-usa-for-39-billion/?hp). *New York Times*. March 20, 2011. . Retrieved March 20, 2011.
[29] Segan, Sascha (2011-03-20). "AT&T Buys T-Mobile: Great For Them, Bad For You" (http://www.pcmag.com/article2/0,2817,2382267,00.asp). Pcmag.com. . Retrieved 2011-11-28.
[30] "BBC.co.uk" (http://www.bbc.co.uk/news/business-12802111). BBC.co.uk. 2011-03-21. . Retrieved 2011-11-28.
[31] German, Kent. "On Call: Civil rights groups line up behind AT&T-T-Mobile merger" (http://www.cnet.com/8301-17918_1-20070630-85/on-call-civil-rights-groups-line-up-behind-at-t-t-mobile-merger/?tag=mncol;txt). *CNET*. . Retrieved 17 August 2011.
[32] German, Kent. "States weigh in on AT&T-T-Mobile merger" (http://www.cnet.com/8301-17918_1-20070911-85/states-weigh-in-on-at-t-t-mobile-merger/?tag=mncol;txt). *CNET*. . Retrieved 17 August 2011.
[33] Sara Jerome (March 23, 2011). "T-Mobile, AT&T merger to draw torrent of opposition". *The Hill*.
[34] Declan McCullagh, CNET. " AT&T defends T-Mobile deal to U.S. Senate (http://news.cnet.com/8301-1035_3-20061889-94.html?part=rss&subj=news&tag=2547-1_3-0-20)." May 11, 2011. Retrieved May 11, 2011.
[35] Don Reisinger, CNET. " Report: AT&T pays $6B if T-Mobile deal fails (http://news.cnet.com/8301-13506_3-20062527-17.html?part=rss&subj=news&tag=2547-1_3-0-20)." May 13, 2011. Retrieved May 13, 2011.
[36] Schoenberg, Tom. "T-Mobile Antitrust Challenge Leaves AT&T With Little Recourse on Takeover" (http://www.bloomberg.com/news/2011-08-31/u-s-files-antitrust-complaint-to-block-proposed-at-t-t-mobile-merger.html). Bloomberg. . Retrieved 2011-11-28.
[37] Goldman, David (2011-09-01). "AT&T and T-Mobile: Is the deal dead? No one knows. - Sep. 1, 2011" (http://money.cnn.com/2011/09/01/technology/att_tmobile_lawsuit/index.htm?hpt=hp_t2). Money.cnn.com. . Retrieved 2011-11-28.
[38] Shields, Todd (2011-11-30). "FCC Allows AT&T to Withdraw Merger Application" (http://www.bloomberg.com/news/2011-11-29/fcc-allows-at-t-to-withdraw-merger-application.html). Bloomberg. . Retrieved 2011-11-30.
[39] AT&T (2011-12-19). "AT&T Ends Bid To Add Network Capacity Through T-Mobile USA Purchase" (http://www.att.com/gen/press-room?pid=22146&cdvn=news&newsarticleid=33560&mapcode=corporate|wireless-networks-general). AT&T. . Retrieved 2011-12-19.
[40] "Top All-Time Donors, 1989-2012" (http://www.opensecrets.org/orgs/list.php?order=A), *OpenSecrets.org* (United States: Center for Responsive Politics), 2011, , retrieved 9 Dec 2011

[41] Kang, Cecelia; Jia Lynn Yang (9 Dec 2011), "How AT&T fumbled its $39 billion bid to acquire T-Mobile" (http://www.washingtonpost.com/business/technology/how-atandt-lost-its-39-million-bid-to-acquire-t-mobile/2011/12/01/gIQAkTQ6hO_story.html), *The Washington Post*: washingtonpost.com, , retrieved 9 Dec 2011
[42] "AT&T Inc: Totals" (http://www.opensecrets.org/orgs/totals.php?cycle=2012&id=D000000076), *OpenSecrets.org* (United States: Center for Responsive Politics), 2011, , retrieved 9 Dec 2011
[43] "AT&T Inc" (http://web.archive.org/web/20070930035728/http://www.opensecrets.org/orgs/summary.asp?ID=D000000076&Name=AT&T+Inc). The Center For Responsive Politics. Archived from the original (http://www.opensecrets.org/orgs/summary.asp?ID=D000000076&Name=AT&T+Inc) on 30 Sep 2007. . Retrieved September 29, 2007.
[44] Drinkard, Jim (January 17, 2005). "Donors get good seats, great access this week" (http://www.usatoday.com/news/washington/2005-01-16-inauguration-donors_x.htm). *USA Today*. . Retrieved May 25, 2008.
[45] "Financing the inauguration" (http://www.usatoday.com/news/washington/2005-01-16-inaugural-donors_x.htm). *USA Today*. January 16, 2005. . Retrieved May 25, 2008.
[46] Associated Press (January 14, 2005). "Some question inaugural's multi-million price tag" (http://www.usatoday.com/news/washington/2005-01-14-price_x.htm). *USA Today*. . Retrieved May 25, 2008.
[47] "Agreements Between SNET America, Inc. (SAI) DBA AT&T Long Distance East, and AT&T Telephone Companies" (http://web.archive.org/web/20071011214558/http://att.com/gen/public-affairs?pid=8101). AT&T. Archived from the original (http://www.att.com/gen/public-affairs?pid=8101) on October 11, 2007. . Retrieved September 29, 2007.
[48] "Current board of directors at AT&T.com" (http://www.att.com/gen/investor-relations?pid=5629). Att.com. . Retrieved 2011-11-28.
[49] "Consumer Reports Names AT&T Worst Wireless Service Provider" (http://techland.time.com/2010/12/06/consumer-reports-names-att-worst-wireless-service-provider/). *Time*. December 6, 2010. . Retrieved December 6, 2010.
[50] Shawn Knight. "AT&T rated worst wireless provider for second year in a row" (http://www.techspot.com/news/46546-att-rated-worst-wireless-provider-for-second-year-in-a-row.html). techspot.com. . Retrieved December 6, 2011.
[51] Grossberg, Josh (August 9, 2007). "AT&T's Pearl Jamming?" (http://www.eonline.com/news/article/index.jsp?uuid=4de684ae-62eb-4b23-984c-d07ce72ea5e2). E Online. . Retrieved September 29, 2007.
[52] Roberts, Michelle (August 10, 2007). "AT&T: Pearl Jam edit a mistake" (http://seattletimes.nwsource.com/html/nationworld/2003829996_pearljam10.html?syndication=rss). *The Seattle Times*. Associated Press. . Retrieved September 29, 2007.
[53] Fisher, Ken (2007-10-01). "AT&T threatens to disconnect subscribers who criticize the company" (http://arstechnica.com/news.ars/post/20070930-att-threatens-to-disconnect-subscribers-who-are-critical-of-the-company.html). Arstechnica.com. . Retrieved 2011-11-28.
[54] "AT&T Legal Policy" (http://home.bellsouth.net/csbellsouth/s/s.dll?spage=cg/legal/att.htm&leg=tos). AT&T. . Retrieved September 29, 2007.
[55] Martin H. Bosworth. "AT&T Changes Terms Of Service After Outcry" (http://www.consumeraffairs.com/news04/2007/10/att_tos.html). Consumeraffairs.com. . Retrieved 2011-11-28.
[56] "AT&T Legal Policy" (http://www.att.net/csbellsouth/s/s.dll?spage=cg/legal/att.htm&leg=tos). Att.net. 2011-05-02. . Retrieved 2011-11-28.
[57] Kincaid, Jason (July 26, 2009). "AT&T Reportedly Blocks 4chan. This Is Going To Get Ugly" (http://www.techcrunch.com/2009/07/26/att-blocks-4chan-this-is-going-to-get-ugly/). TechCrunch. . Retrieved July 27, 2009.
[58] Cheng, Jacqui (July 27, 2009). "AT&T: 4chan block due to DDoS attack coming from 4chan IPs" (http://arstechnica.com/telecom/news/2009/07/att-4chan-block-due-to-ddos-attack-coming-from-4chan-ips.ars). Ars Technica. . Retrieved July 28, 2009.
[59] Feared Hackers Call Off Attack on AT&T. Tuesday, July 28, 2009. (http://www.foxnews.com/story/0,2933,534941,00.html) Retrieved July 28, 2009
[60] When Your Pedicurist Is A Fish (dated July 22, 2008) (http://www.npr.org/templates/story/story.php?storyId=92767742) Transcript of National Public Radio news interview. Retrieved July 28, 2009
[61] "Klein Exhibit" Document from Hepting vs AT&T lawsuit from 2007. Reported by Ryan Singel in Wired Magazine, article "AT&T 'Spy Room' Documents Unsealed; You've Already Seen Them" (http://www.wired.com/politics/law/news/2007/06/spy_room) June 13, 2007, Documents posted at the Electronic Frontier Foundation (http://www.eff.org) website: (File "SER_klein_exhibits.pdf") (http://eff.org/legal/cases/att)
[62] Nakashima, Ellen, "A Story of Surveillance" (http://www.washingtonpost.com/wp-dyn/content/article/2007/11/07/AR2007110700006_pf.html), Washington Post, November 7, 2007
[63] Singel, Ryan (April 7, 2006). "Whistle-Blower Outs NSA Spy Room" (http://www.wired.com/science/discoveries/news/2006/04/70619). Wired. . Retrieved September 29, 2007.
[64] "Government Moves to Intervene in AT&T Surveillance Case" (http://www.eff.org/news/archives/2006_04.php#004613) (Press release). Electronic Frontier Foundation (EFF). April 28, 2006. . Retrieved September 29, 2007.
[65] U-verse
[66] Cauley, Leslie (May 11, 2006). "NSA has massive database of Americans' phone calls" (http://www.usatoday.com/news/washington/2006-05-10-nsa_x.htm). *USA Today*. . Retrieved September 29, 2007.
[67] Lazarus, David (June 21, 2006). "AT&T Rewrites Rules: Your Data Isn't Yours" (http://www.sfgate.com/cgi-bin/article.cgi?f=/c/a/2006/06/21/BUG9VJHB9C1.DTL&hw=at&sn=002&sc=870). *San Francisco Chronicle*. . Retrieved September 29, 2007.
[68] Shrader, Katherine (August 22, 2007). "Spy Chief Reveals Classified Surveillance Details" (http://www.msnbc.msn.com/id/20396282/). Associated Press. . Retrieved September 29, 2007.

[69] Olbermann, Keith (November 8, 2007). "Whistleblower Saw AT&T Assist Bush Administration" (http://video.msn.com/video.aspx?mkt=en-US&brand=msnbc&vid=297abdd5-d0dc-4617-a6c9-c482fa316b59). MSNBC. . Retrieved November 10, 2007.
[70] "Document Shows How Phone Cos. Treat Private Data" (http://www.npr.org/templates/story/story.php?storyId=140931876). Associated Press. September 29, 2011. . Retrieved 2011-09-29. "T-Mobile USA doesn't keep any information on Web browsing activity. Verizon, on the other hand, keeps some information for up to a year that can be used to ascertain if a particular phone visited a particular Web site. According to the sheet, Sprint Nextel Corp.'s Virgin Mobile brand keeps the text content of text messages for three months. Verizon keeps it for three to five days. None of the other carriers keep texts at all, but they keep records of who texted who for more than a year. The document says AT&T keeps for five to seven years a record of who text messages who —and when, but not the content of the messages. Virgin Mobile only keeps that data for two to three months."
[71] Wu, Tim (January 16, 2008). "Has AT&T Lost Its Mind? A baffling proposal to filter the Internet" (http://www.slate.com/id/2182152/). Slate. .
[72] "AT&T's Proposed Filtering Policy Is Bad News – Netiquette – MSNBC.com" (http://www.msnbc.msn.com/id/22829568/). MSNBC. 2008-01-25. . Retrieved 2011-11-28.
[73] (March 9, 2009) "AT&T Accused of Discriminating Against Local Public Access Channels, Deadline for Public Comment Expires Tonight" (http://www.democracynow.org/2009/3/9/at_t_accused_of_discriminating_against), Democracy Now!, retrieved on March 13, 2009.
[74] Keizer, Gregg (June 10, 2010). "'Brute force' script snatched iPad e-mail addresses" (http://www.computerworld.com/s/article/9177921/_Brute_force_script_snatched_iPad_e_mail_addresses). *Computerworld*. . Retrieved September 18, 2010.
[75] Keizer, Gregg (June 11, 2010). "iPad e-mail hackers defend attack as 'ethical'" (http://www.computerworld.com/s/article/9177991/iPad_e_mail_hackers_defend_attack_as_ethical_?taxonomyId=17&pageNumber=2). *Computerworld*: p. 2. . Retrieved September 25, 2010.
[76] Keizer, Gregg (June 11, 2010). "iPad e-mail hackers defend attack as 'ethical'" (http://www.computerworld.com/s/article/9177991/iPad_e_mail_hackers_defend_attack_as_ethical_). *Computerworld*: p. 1. . Retrieved September 25, 2010.
[77] "Teamusa.org" (http://www.teamusa.org/pages/sponsors). Teamusa.org. . Retrieved 2011-11-28.
[78] "NCAA.org" (http://ncaa.org/wps/portal/ncaahome?WCM_GLOBAL_CONTEXT=/corp_relations/CorpRel/Corporate+Relationships/Corporate+Alliances/partners.html). NCAA.org. 2007-12-14. . Retrieved 2011-11-28.
[79] "Corp.att.com" (http://www.corp.att.com/ap/about/where/hk/). Corp.att.com. . Retrieved 2011-11-28.
[80] Matthew Nickasch, Network World. " Toshiba Announces AT&T SIP Trunking Interoperability (http://www.networkworld.com/community/node/31208)." August 19, 2008. Retrieved October 12, 2011.
[81] Paul Korzeniowski, Information Week. " AT&T VPN Service Gains Its Voice (http://www.informationweek.com/blog/smb/229200865)." August 24, 2010. Retrieved October 12, 2011.

External links

Corporate information

- Official website (http://http://www.att.com)
- Brand evolution of AT&T companies (http://www.att.com/Common/attrev1/331012_timeline_evolution16.pdf)
- AT&T (http://www.fi.edu/learn/case-files/att/) History and science resources at The Franklin Institute's Case Files online exhibit
- Press Release announcing FCC Approval of SBC-Ameritech merger (http://www.fcc.gov/Bureaus/Common_Carrier/News_Releases/1999/nrc9077a.html) (October 6, 1999)

Articles

- Thierer, Adam D.. "Unnatural Monopoly: Critical Moments in the Development of the Bell System Monopoly" (http://www.cato.org/pubs/journal/cjv14n2-6.html). cato.org.
- "AT&T buys IBM's Global Network" (http://news.bbc.co.uk/1/hi/business/the_company_file/230831.stm). *BBC News*. December 8, 1998.
- Reardon, Marguerite (November 18, 2005). "SBC closes AT&T acquisition" (http://news.cnet.com/SBC+closes+AT38T+acquisition/2100-1036_3-5961206.html). CNet News.
- Reardon, Marguerite (March 5, 2006). "AT&T to buy BellSouth for $67 billion" (http://news.cnet.com/AT38T+to+buy+BellSouth+for+67+billion/2100-1037_3-6046081.html). CNet News.
- "AT&T Whistleblower to Urge Senate to Reject Blanket Immunity for Telecoms" (http://www.eff.org/press/archives/2007/11/05). Electronic Frontier Foundation.

Samsung_Galaxy_S_II

GALAXY S II	
 Samsung Galaxy S II (GT-I9100)	
Manufacturer	Samsung Electronics
Series	S series
Compatible networks	Dual band CDMA2000/EV-DO Rev. A 800 and 1900 MHz; WiMAX 2.5 to 2.7 GHz; 802.16e 2.5G (GSM/GPRS/EDGE): 850, 900, 1800, and 1900 MHz UMTS: 850, 900, 1700 (T-Mobile USA only), 1900, and 2100 MHz HSPA+: 21/42 Mbit/s; HSUPA: 5.76 Mbit/s LTE 700/1700 Rogers Only
Predecessor	Samsung Galaxy S
Successor	Samsung Galaxy S III
Related	Galaxy Note Galaxy Ace Galaxy Nexus Infuse 4G
Type	Touchscreen smartphone
Dimensions	125.3 mm (4.93 in) H 66.1 mm (2.60 in) W 8.49 mm (0.334 in) D (standard) 5.11 in (130 mm) H 2.74 in (70 mm) W 0.38 in (9.7 mm) D (Sprint)
Weight	4.6 oz (130 g) (standard) 116 g (4.1 oz) (Sprint)
Operating system	Android 2.3.6 (Gingerbread) with TouchWiz UI 4.0
CPU	1.2 GHz dual-core ARM Cortex-A9[1] SoC processor; Samsung Exynos (GT-I9100)(under AT&T, Sprint, and outside US models); Texas Instruments OMAP4430 (GT-I9100G) 1.5 GHz dual core Qualcomm Snapdragon S3 (T-Mobile Model / SGSII LTE & HD LTE Model & Rogers Model)
GPU	ARM Mali-400 MP (GT-I9100)[2] [3] Adreno 220 (models with Snapdragon SoC) PowerVR SGX540 (GT-I9100G)
Memory	1 GB RAM

Storage	16 GB flash memory
Removable storage	microSD (up to 32 GB)
Battery	Li-ion 1850mAh Rogers Model & 1650 mAh
Data inputs	Multi-touch touch screen, headset controls, proximity and ambient light sensors, 3-axis gyroscope, magnetometer, accelerometer, aGPS, and stereo FM-radio[4]
Display	800×480 px, 10.8 cm (4.3 in) at 218 ppi WVGA Super AMOLED Plus (0.37 megapixels) 4.5" Rogers Model Only
Rear camera	8 Mpx Back-illuminated sensor with auto focus, 1080p 30 fps full HD video recording, full HD reproduction, and stills. Single LED flash.
Front camera	2 Mpx for video chatting, video recording (VGA), and stills
Connectivity	3.5 mm TRRS; Wi-Fi (802.11a/b/g/n); Wi-Fi Direct; Bluetooth 3.0; micro USB 2.0; Near field communication (NFC); DLNA; MHL; HDMI; USB Host (OTG) 2.0
Other	Exchange ActiveSync, integrated messaging *Social Hub*, *Readers Hub*, *Music Hub*, and *Game Hub*
Hearing aid compatibility	M3/T3[5]

The **Samsung Galaxy S II** (GT-I9100) is a smartphone running under the Android operating system that was announced by Samsung on February 13, 2011 at the Mobile World Congress. It is the successor to the Samsung Galaxy S, with a different appearance and significantly improved hardware.[6] The Galaxy S II was one of the slimmest smartphones of the time, mostly 8.49 mm thick, except for two small bulges which take the total thickness of the phone to 9.91 mm.[7]

The Galaxy S II has a 1.2 GHz dual-core "Exynos" system on a chip (SoC) processor,[8] 1 GB of RAM, a 10.8 cm (4.3 in) WVGA Super AMOLED Plus screen display and an 8 megapixel camera with flash and full 1080p high definition video recording. It is one of the first devices to offer a Mobile High-definition Link (MHL),[9] which allows up to 1080p uncompressed video output with HDMI while charging the device at the same time. USB On-The-Go (USB OTG) is supported.[10] [11]

The user-replaceable battery on the Galaxy S II gives up to ten hours of heavy usage, or two days of lighter usage.[12] According to Samsung, the Galaxy S II is capable of providing 9 hours of talk time on 3G and 18.3 hours on 2G.[12] [13]

Launch and availability

The Galaxy S II was given worldwide release dates starting from May 2011, by more than 140 vendors in some 120 countries.[14]

On May 9, 2011, Samsung announced that they had received pre-orders for 3 million Galaxy S II units globally.[15]

In some time after the device's release, Samsung also released a Nvidia Tegra 2 powered Samsung Galaxy R (or 'Galaxy Z' in Sweden) variant version. The release of the Galaxy R 'GT-I9103' corroborated with early pre-release reports of an Nvidia powered Galaxy S II smartphone.[16] [17] Eldar Murtazin, of Mobile-Review.com, hinted that Samsung might have faced delays in meeting worldwide shipment of both its Exynos chip and Super AMOLED Plus screens. He noted that nobody expected the "huge success" and "sky high" demand for the previous Samsung Galaxy S.[18]

Different regions around the world have received this device to market beginning from the earliest of May 2011.

Asia/Pacific

Country	Release date
Australia	June 1, 2011
Hong Kong	June 14, 2011
Israel	July 3, 2011
India	June 8, 2011[19]
Iran	June 1, 2011
Iraq	July 18, 2011
Japan	June 23, 2011
Lebanon	June 21, 2011
Malaysia	June 22, 2011
Pakistan	July 23, 2011[20]
Philippines	June 27, 2011
Singapore	June 24, 2011
South Korea	April 29, 2011
Taiwan	June 17, 2011
Thailand	July 8, 2011
UAE	June 22, 2011

North America

Country	Carrier	Release date	Variant
Canada	Bell Mobility, Virgin Mobile Canada	July, 2011[21] [22]	Samsung Galaxy S II 4G
Canada	Telus	October, 2011	Samsung Galaxy S II X
Canada	Rogers Wireless	November, 2011	Samsung Galaxy S II LTE (i727R)
Mexico	Telcel	June, 2011[23]	Samsung Galaxy II LTE
United States	Sprint	September 16, 2011	Samsung 'EPIC 4G Touch'[24]
United States	AT&T	November 6, 2011	Samsung Galaxy S II Skyrocket i727
United States	T-Mobile	October 12, 2011	Samsung Galaxy S II T-Mobile
Caribbean	Digicel	December 21, 2011	Samsung Galaxy S II Digicel

South America

Country	Release date
Chile	May 25, 2011
Brazil	June 28, 2011
Argentina	August 22, 2011

Africa

Country	Release date
South Africa	June, 2011[25]
Uganda	September, 2011

Europe

Country	Release date
Austria	May 5, 2011
Denmark	May 19, 2011
Belgium	June 1, 2011
Finland	May 25, 2011
France	June 23, 2011
Germany	May 16, 2011
Hungary	June 1, 2011
Ireland	June 1, 2011
Italy	May 25, 2011
Netherlands	May 11, 2011
Norway	May 26, 2011
Portugal	June 8, 2011
Romania	May 26, 2011
Russia	May 18, 2011
Spain	May 15, 2011
Sweden	May 12, 2011
Turkey	June 9, 2011
UK	May 1, 2011

Hardware

Processor

The Galaxy S II has a 1.2 GHz dual core ARM Cortex-A9 processor that uses Samsung's own 'Exynos 4210' System on a chip (SoC) that was previously code-named "Orion".

Dismantled Samsung Galaxy S II, from left to right components include the handset, battery and back cover

The Exynos branded SoC was the source of much speculation concerning another branded successor to the previous "Hummingbird" single-core SoC of the Samsung Galaxy S. The Exynos 4210 uses ARM's Mali-400 MP GPU.[2] [26] This graphics GPU, supplied by ARM, is a move away from the PowerVR GPU of the Samsung Galaxy S.[27]

The Exynos 4210 supports ARM's SIMD engine (also known as *Media Processing Engine*, or 'NEON' instructions), and may give a significant performance advantage in critical performance situations such as accelerated decoding for many multimedia codecs and formats (e.g., On2's VP6/7/8 or Real formats).[28] [29] [30]

At the 2011 Game Developers Conference ARM's representatives demonstrated 60 Hz framerate playback in stereoscopic 3D running on the same Mali-400 MP and Exynos SoC. They said that an increased framerate of 70 Hz would be possible through the use of an HDMI 1.4 port.[26]

The Motorola Atrix advertised in June 2011 that it was "the world's most powerful smartphone"; in August 2011 the UK Advertising Standards Authority ruled that the Atrix was not as powerful as Galaxy SII due to its faster processor.[31]

A newer Samsung Galaxy S II (i9100G) uses a 1.2 Ghz dual core TI OMAP 4430 processor with PowerVR SGX540 graphics.[32]

Memory

The Galaxy S II has 1 GB of dedicated RAM (in either LPDDR or possibly DDR2/DDR3 by Samsung) and 16 GB of internal mass storage. Within the battery compartment there is an external microSD card slot.

Display

The Samsung Galaxy S II uses a 108.5-millimetre (4.27 in)[11] WVGA (800 x 480) Super AMOLED Plus capacitive touchscreen that is covered by Gorilla Glass with an oleophobic fingerprint-resistant coating. The display is an upgrade of its predecessor, and the "Plus" signifies that the display panel has done away with Pentile matrix to regular RGB matrix display which results in a 50% increase in sub-pixels. This translates to grain reduction and sharper images and text. In addition, Samsung has claimed that Super AMOLED Plus displays are 18% more power efficient than the older Super AMOLED displays.[33] Some phones have display issues, with a few users reporting a "yellow tint" on the left bottom edge of the display when a neutral grey background is displayed.[34]

Samsung Galaxy S II with Super AMOLED Plus display

Audio

The Galaxy S II uses Yamaha audio hardware.[35] The Galaxy S II's predecessor, the original Galaxy S, used Wolfson's WM8994 DAC.[36] User feedback on Internet forums as well as an in-depth review at *Clove*,[35] have expressed the Yamaha chip's inferior sound quality compared to that of the Wolfson chip featured in the original Galaxy S.

Camera

On the back of the device is an 8-megapixel Back-illuminated sensor[37] camera with single-LED flash that can record videos in full high-definition 1080p at 30 frames per second. There is also a fixed focus front-facing 2-megapixel camera for video calling, taking photos as well as general video recording, with a maximum resolution of VGA (640×480).

Connectivity

The Galaxy S II is one of the earliest Android devices to natively support near field communication (NFC).[38] This follows on from the Google Nexus S which was the first de-facto NFC smartphone device.[39] It has been reported that the UK version will be supplied without an NFC chip at the beginning of its production run,[40] with an NFC-equipped version released later in 2011.[41]

Samsung has also included a new high-definition connection technology called Mobile High-definition Link (MHL). The main specialty of MHL is that it is optimized for mobile devices by allowing the device's battery to be charged while at the same time playing back multimedia content.[42] For the Galaxy S II, the industry standard micro USB port found on the bottom of the device can be used with an MHL connector for a TV out connection to an external display, such as a high definition television.[43]

The micro USB port on this device also supports USB On-The-Go (USB OTG) standard which means the Galaxy S II can act as a 'host' device in the same way as a desktop computer in allowing external USB devices to be plugged in and used.[10] These external USB devices typically include USB flash drives and separately powered external hard drives. A video demonstration on YouTube[44] has shown the OTG function to be readily available with an ordinary micro USB (B-type) OTG adaptor. The same YouTube video goes on to mention a successful test completed on a 2 TB USB external hard drive (requiring own power source) but however reports of failure when trying to connect USB keyboards, tested USB mice and tested USB game pads. Currently the only file-system supported for USB drives within OTG is Fat32.

A 3.5 mm TRRS headset jack is available and is located on the top side of the device. The micro USB connection port is located on the bottom side of the device.

BCM4330 combo chip integrates 802.11n Wi-Fi, Bluetooth 3.0 + HS and FM radio. BCM4330 supports Wi-Fi Direct that communicate directly with one another without having to interact with an access point.[45]

Accessories (optional)

- Dock connector for battery charging and audio-visual output[46]
- MHL cable which makes use of the device's micro USB port for HDMI output[46]
- USB OTG adaptor for use with external USB devices such as USB flash drives.[10]
- Stylus pen for use on the device's capacitive screen.[46] Support for a stylus on the Galaxy S II is precursor to the Samsung Galaxy Note.

Software

Android operating system

The Galaxy S II ships with Android's 2.3 (Gingerbread) installed.[4]

Samsung Galaxy S II US variants begun shipments with the slightly updated Android 2.3.5 (Gingerbread) installed.[47] [48] The Android 2.3.6 firmware update from Samsung was made globally available from Dec 12 2011.[49]

In December 2011, Samsung confirmed that it would offer a firmware update to Android 4.0 (Ice Cream Sandwich) within the first quarter of 2012.

User interface

The phone employs the latest proprietary Samsung TouchWiz 4.0 user interface. It follows the same principle as TouchWiz 3.0 found on the previous Galaxy but adds new improvements, such as hardware acceleration. It also has a new optional gesture based interaction called 'motion' which (among other things) allows users to zoom in and out by placing two fingers on the screen and tilting the device towards and away from you to zoom in and out respectively. This gesture function works on both the web browser and the images in gallery[50] used within this device. There have been improvements to the widgets drawer and layout in how many widgets can be added and how they are presented. Additionally there is another new optional gesture-based control called 'panning' on TouchWiz 4.0 for the movement of widgets and icons shortcuts between screens, by allowing the device to be held and moved from side to side to scroll through home screens. This gesture-based management of widgets is a new optional method next to the existing method of holding and swiping between home screens.[51]

Bundled applications

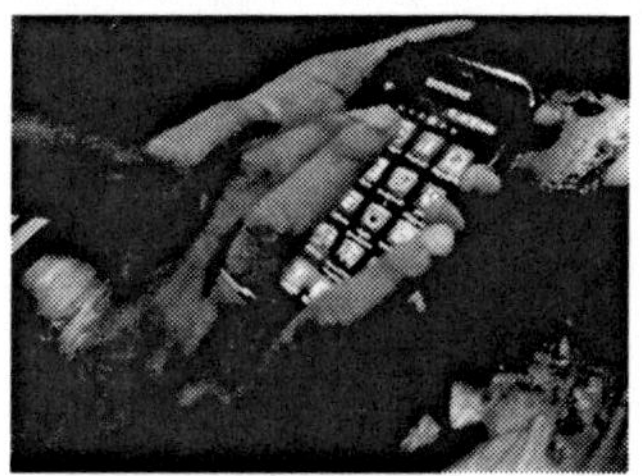
Galaxy with several applications

Four new Samsung 'Hub' applications were revealed at the 2011 Mobile World Congress:

Social Hub

Which integrates popular social networking services like Facebook and Twitter into one place rather than in separate applications.

Readers Hub

This hub provides the ability to access, read and download online newspapers, ebooks and magazines from a worldwide selection.

Music Hub

An application store for downloading and purchasing music tracks on the device. Samsung has teamed up with 7digital to offer this service.[52]

Game Hub

An application store for downloading and purchasing games. Samsung has teamed up with partners including Gameloft to offer this service.[52]

Other applications

More applications include Kies 2.0, Kies Air,[53] AllShare (for DLNA), Voice Recognition, Google Voice Translation,[54] Google Maps with Latitude, Places, Navigation (beta) and Lost Phone Management, Adobe Flash 10.2, QuickOffice application and 'QuickType' by SWYPE.

Before launch, it was announced that Samsung had taken steps to incorporate Enterprise software for business users, which included On Device Encryption, Cisco's AnyConnect VPN, MDM (Mobile Device Management), Cisco WebEx, Juniper,[38] and secure remote device management from Sybase.[55]

Cisco's AnyConnect VPN for Samsung devices is now currently available on Android Market,.[56]

Media support

The Galaxy S II comes with support for many multimedia file formats and codecs. For audio it supports FLAC, WAV, Vorbis, MP3, AAC, AAC+, eAAC+, WMA, AMR-NB, AMR-WB, MID, AC3, XMF. For video formats and codecs it supports MPEG-4, H.264, H.263, DivX HD/XviD, VC-1, 3GP (MPEG-4), WMV (ASF) as well as AVI (DivX)), MKV, FLV and the Sorenson codec. For H.264 playback, the device natively supports both 8-bit and 10-bit encodes along with up to 1080p HD video playback.[57] [58] [59]

Community support

It was public news that Samsung sent a number of Galaxy S II devices to four developers of the CyanogenMod project, with the only request from Samsung being to bring full support of CyanogenMod to the device.[60] [61]

Reception

Reviews of the Galaxy S II have been positive. *Engadget* gave the device a 9/10, calling it "the best Android smartphone yet" and "possibly the best smartphone, period."[62] *CNET* UK gave the device a favorable review of 4.5/5 and described it as "one of the slimmest, lightest mobiles we've ever had the privilege to hold."[63] *TechRadar* gave the device 5/5 stars and describes the device as one that "set a new bar for smartphones in 2011."[64] *Pocketnow* was "impressed" with the speed of the web browser.[65] *SlashGear* states that the device "sets the benchmark for smartphones in general."[66] *GSMArena* points out minor drawbacks such as an "all-plastic body" and the handset having "no dedicated camera key," but still calls the handset "absurdly powerful" and concluding "we just cannot see beyond the new Samsung flagship if we're to name the ultimate smartphone."[67]

After slightly over one month since its debut, more than 1 million units of Samsung Galaxy S II were activated in South Korea.[68] Worldwide, 3 million units were sold in 55 days.[69] After over 85 days of its first release, Samsung has declared global shipments of over 5 million for Galaxy S II[70] and 10 million after 5 months.[71] Partially owing to strong sales of Samsung's Galaxy range of smartphones, Samsung overtook Apple in smartphone sales during Q3 2011, with a total market share of 23.8%, compared to Apple's 14.6%.[72]

Variants

Galaxy R (GT-I9103)

The Samsung Galaxy R is one of the currently available variants of the Galaxy S II. It was announced for releases first in Europe starting from the end of July 2011 with subsequent releases for worldwide. The Galaxy R is slightly smaller in physical size along with some notable "downgraded" features.[73] [74] Its differing comparable features include a 10.2 cm (4.0 in) SC-LCD capacitive touchscreen display, a 5 megapixel camera with 720p HD video recording, 16GB internal storage and a microSD memory card slot. The overall physical size and back of the design of this device differ slightly to that of the original Galaxy S II. Like Galaxy S II, the Galaxy R will also support Kies Air (PC Suite via Wi-Fi).[16] [17]

Galaxy W (Galaxy S II Mini)

The Samsung Galaxy W I8150 is a smaller-sized 9.5 cm (3.7 in) touchscreen variant. The Galaxy W, too, is a downgrade in features to that of the Galaxy S II with the same comparable downgraded features as that of the Galaxy R. The Galaxy W and Galaxy R both feature a 'SC-LCD' capacitive touchscreen with continued 480x800 resolution, both Galaxy W and Galaxy R feature a 5 megapixel still-image camera with 720p video capture. The main differences of the Galaxy W from other variants, is that it features a 1.4 GHz Qualcomm manufactured processor, its physical design differs slightly and furthermore its screen has a higher pixel density (ppi) compared to that of the Galaxy S II and Galaxy R.[75] [76]

Prior to the release of the Galaxy S II, there were speculative reports of an exact same smaller-sized, stripped down variant planned for release; in a similar example to the way the HTC HD Mini was towards the HTC HD2. The name previously rumored for this smaller Galaxy S II variant included the naming suffix of 'Mini' added the end of its name.[77] [78]

Galaxy S II LTE and Galaxy S II HD LTE

Announced on August 28, 2011, the Galaxy S II *Long Term Evolution* (LTE) and Galaxy S II HD (LTE) mobile phones are part of the Galaxy S II family. They run on the Android 2.3 Gingerbread OS, have 16GB of internal memory, and feature a 1.5 GHz dual-core processor. The S II LTE has a 4.5-inch wide Super AMOLED display. The S II HD LTE features a 4.65-inch high-definition AMOLED display, offering a screen of 720x1280 pixels – 720p HD resolution.[79] Both of these variants have NFC enabled.[80]

U.S. variants

Sprint's Galaxy S II Epic 4G Touch

The Sprint version of the device, previously codenamed "Within", is known as the Samsung Galaxy S II Epic 4G Touch (Model SPH-D710). Unlike its predecessor, the Epic 4G, the Epic 4G Touch lacks a physical QWERTY keyboard. At 4.52 inches, it has a larger display than the international version.[81] Additionally, the Epic 4G Touch features four touch-capacitive buttons, as opposed to the hardware/capacitive button combination found on the international version. Other differences include an LED notification light, larger battery (6.66Wh), and a WiMax radio. The device was launched on September 16, 2011.[82] Unlike the T-Mobile USA and AT&T versions this particular model does not come with the NFC chip.[83]

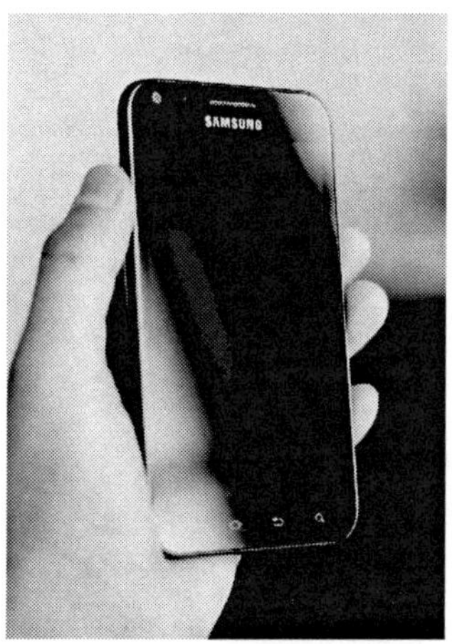

Sprint's Variant: Samsung Galaxy S II Epic 4G Touch

T-Mobile's Galaxy S II

The T-Mobile USA version was codenamed the Samsung "Hercules" and given the model number **SGH-T989** (as opposed to other phones with the Galaxy S II name which are I9100 models). It has a 4.52 inch screen, and features four touch buttons.[84] The cellular radio in this variant supports UMTS Band I (2100), Band II (1900), Band IV (1700), and Band V (850). It also contains NFC.[85] It uses a 1.5 GHz dual-core Qualcomm APQ8060 (S3) Snapdragon processor, as opposed to the 1.2Ghz dual-core Exynos processor. The reason for this is that the Exynos processor is not compatible with T-Mobile's 42 Mbit/s HSPA+ network. As opposed to most of their other phones, T-Mobile's version was initially released with no support for Wi-Fi calling for calls on their own network, but Wi-Fi calling through third party apps such as Skype is possible. T-Mobile and Samsung released an update for Wi-Fi calling on December 15, 2011. The update is available from Samsung via KIES for the T989 and OTA (unconfirmed) from T Mobile. The S II was available for pre-order through T-Mobile's website on October 10, 2011 and became available in stores on October 12.[86]

AT&T's Galaxy S II

The AT&T version was codenamed "Attain" (Model SGH-I777). However, it appeared on AT&T and Samsung websites simply as the Samsung Galaxy S II. It maintains a 4.3 inch screen, but also features four touch buttons.[48] It also contains NFC.[85] It was released on October 2, 2011.

AT&T's Galaxy S II: Skyrocket

This is the second variant of this phone to come to AT&T (Model SGH-I727) along with some changes. For instance, the display features the same 4.52 inch screen shared with the Sprint model. Additionally, this variant also includes LTE radio, to coincide with AT&T's launch of its LTE network. However, to make this change it was required for the SoC chipset to be changed from the Exynos to the Qualcomm Snapdragon MSM8260, as the Exynos SoC does not support LTE (akin to a similar limitation made on the T-Mobile model). 'Skyrocket' was released on November 6, 2011 alongside a similar LTE-capable [HTC Vivid] at the same date. The NFC feature is not enabled.

AT&T announced at its CES 2012 Developer Summit keynote that the Galaxy S II Skyrocket HD would be released early in 2012. The Skyrocket HD will be the US variant of the Galaxy S II LTE HD and will feature a 4.65" Super AMOLED Plus display that is capable of 1280x720p resolution, the same 1.5gHz processor as the Skyrocket, and an LTE radio. No release date was specified.

AT&T's Samsung Captivate Glide

This is the third variant to come out on AT&T (Model SGH-I927)[87] It includes a dual-core processor, but is a 1 GHz Tegra 2 processor instead of the 1.2 GHz Exynos processor. Its display is Super AMOLED is instead of Super AMOLED Plus, and the display size is reduced to 4 inches. It can record video in 720p resolution at 30 fps. Its biggest attraction is the addition of a slide-out physical QWERTY keyboard.

Japanese model (NTT DoCoMo)

In Japan, the Galaxy S2 is known as the NTT DoCoMo SC-02C, manufactured by Samsung Electronics and branded by the DoCoMo carrier, and is the successor to the DoCoMo SC-02B (Galaxy S) within the DoCoMo Smartphone range. Released on 23 June 2011, the Japanese model includes additional features such as 1seg terrestrial television support, as well as i-mode software functions specific to DoCoMo handsets, such as i-channel, BeeTV, MelodyCall and DoCoMo map navigation. The SC-02C uses the Wnn Japanese input system, and uses the Samsung Exynos 4210 Orion Dual-core 1.2 GHz (S5PC210) processor.

South Korean variants

All of south Korean variants have a T-DMB tuner in place of FM radio tuner. They also have Samsung's Korean input system for feature phones as well as Dubeolsik layout in their virtual keyboard.

SK Telecom's Galaxy S II (SHW-M250S)

The SK Telecom variant uses SK-MMS system instead of OMA-MMS system for MMS. Additional features for SK Telecom users are installed by default.

KT's Galaxy S II (SHW-M250K)

This variant uses KT's Wi-Fi CM instead of Android's Wi-Fi CM to connect Wi-Fi. Additional features for KT users are installed by default.

LG U+'s Galaxy S II (SHW-M250L)

Because of mobile networks LG U+ have, it uses EV-DO Rev.B (KPCS 1.8Ghz) instead of WCDMA and HSPA. It is slightly thicker (9.4mm) than SK Telecom and KT variants (8.89mm). Additional features for LG U+ users are installed by default.

Canadian variants

Rogers' Galaxy S II LTE (SGH-i727R)

This phone features a larger screen 4.5", a bigger battery 1850mAh, and a different 1.5GHZ Qualcomm processor. Rogers will carry the Samsung Galaxy S II LTE, launching in Fall 2011, soon after its LTE Launch in Toronto.[88]

Bell's Galaxy S II (GT-I9100M)

Bell's Samsung Galaxy S II is identical to the international version except that its model number is I9100M.

Telus' 4G Galaxy S II X (SGH-T989D)

Telus' Galaxy S II is virtually identical to the T-Mobile variant. It has a Qualcomm 1.5 Ghz dual core processor, larger 4.52 inch screen and 1850 mAh battery, is thicker at 9.4mm and has a different design.[89] There is a chrome band around the edge and the plastic on the back has a leathery feel.[90] Instead of the hardware home button, it has the standard four capacitive buttons. The Qualcomm processor allows for 42Mbps HSPA+ download speeds that the Samsung Exynos processor is not currently capable of. It was released on October 28, 2011.

Successor

JK Shin, the head of "Samsung Mobile Communications", announced in May 2011 that a Samsung Galaxy S III (Galaxy S3) is planned for release in 2012.[91] [92]

See also

- List of Android devices
- Comparison of smartphones
- Comparison of Android devices
- LG Optimus 2X – similar CPU and GPU
- Motorola Atrix 4G – similar CPU and GPU
- HTC Sensation – similar CPU and GPU
- Samsung Infuse 4G
- Galaxy Nexus

References

[1] "Samsung announces UK availability of the GALAXY S II" (http://www.samsung.com/uk/news/newsRead.do?news_seq=27412&gltype=localnews). *Samsung UK*. 13 April 2011. .

[2] "Mali-400 MP – ARM" (http://www.arm.com/products/multimedia/mali-graphics-hardware/mali-400-mp.php). *ARM.com*. . Retrieved 6 August 2011.

[3] "Samsung's Galaxy S II Preliminary Performance: Mali-400MP Benchmarked" (http://www.anandtech.com/show/4177/samsungs-galaxy-s-ii-preliminary-performance-mali400-benchmarked). *Anandtech*. February 14, 2011. .

[4] "Samsung Announces the GALAXY S II, World's Thinnest Smartphone that Will Let You Experience More with Less" (http://www.samsung.com/au/news/newsPreviewRead.do?news_seq=24298). *Samsung.com*. 15 February 2011. .

[5] "Sprint Relay Store" (http://www.sprintrelaystore.com/). . Retrieved 2011-12-15.

[6] "Samsung Galaxy S II gets official at MWC 2011" (http://www.newsden.net/samsung-galaxy-sii-s2-officially-announced-hand-on-video-6607/). .

[7] Ionescu, Daniel. (2011-09-14) Article stating measured dimensions (http://www.pcworld.com/article/240075/iphone_4_worlds_thinnest_phone_not_samsung_galaxy_s_ii.html). Pcworld.com. Retrieved on 2011-11-28.

[8] "Samsung Exynos 4210" (http://www.samsung.com/global/business/semiconductor/products/mobilesoc/Exynos/products4210.html). *Samsung Semiconductor*. . Retrieved 29 September 2011.

[9] "Mobile High-Definition Link (MHL)" (http://armdevices.net/2011/02/15/mobile-high-definition-link-mhl/). *ARMdevices.net*. February 15, 2011. .

[10] "This is how you plug your USB flash drive into the Samsung Galaxy S II" (http://blog.gsmarena.com/this-is-how-you-plug-your-usb-flash-drive-into-the-samsung-galaxy-s-ii/). *GSMArena Blog*. May 12, 2011. .

[11] "Samsung GALAXY S II Specification" (http://www.samsung.com/global/microsite/galaxys2/html/specification.html). *Samsung.com*. . Retrieved April 21, 2011.

[12] Samsung Galaxy S2 review: Battery life and connectivity (http://www.techradar.com/reviews/phones/mobile-phones/samsung-galaxy-s2-930907/review?artc_pg=9). Techradar.com (2011-04-26). Retrieved on 2011-11-28.

[13] Samsung Galaxy S II specs (http://www.phonearena.com/phones/Samsung-Galaxy-S-II_id5106). Phonearena.com. Retrieved on 2011-11-28.

[14] "Samsung challenges Apple with new smartphone" (http://www.google.com/hostednews/afp/article/ALeqM5jifPR0ObtyXNiRtsEN-1vv4BCVyg?docId=CNG.329c6cccb28bef19304d6977b2e73a97.601). *AFP*. April 28, 2011. .

[15] "3m Galaxy S II units are pre-ordered by carriers globally" (http://www.samsunghub.com/2011/05/09/3m-galaxy-s-ii-units-are-pre-ordered-by-carriers-globally/). *Samsung Hub*. 9 May 2011. .

[16] "Samsung I9103 Galaxy R preview: First look - GSMArena.com" (http://www.gsmarena.com/samsung_i9103_galaxy_z-review-614.php). *GSMArena.com*. 30 June 2011. . Retrieved 20 August 2011.

[17] "Samsung Galaxy R official – NVIDIA Tegra 2-toting Galaxy S 2 variant" (http://www.intomobile.com/2011/08/10/samsung-galaxy-r-official-nvidia-tegra-2-dualcore-cpu-42-inch-display-5-megapixel-camera-goodness/). *IntoMobile*. 10 August 2011. . Retrieved 20 August 2011.

[18] Preview of Samsung Galaxy S II i9100. Part 1 (http://www.mobile-review.com/review/samsung-galaxy-s2-en.shtml#2). Mobile-review.com (2011-05-05). Retrieved on 2011-11-28.

[19] Samsung Launches Galaxy S2 in India < Mobile Phones, Samsung, Mobile Phones < PC World India News < PC World.in (http://www.pcworld.in/news/samsung-launches-galaxy-s2-india-49632011). Pcworld.in (2011-05-25). Retrieved on 2011-11-28.

[20] Samsung Galaxy S II Now Available in (Pakistan) Teletec Stores for Rs. 57,000 (http://www.androidpakistan.com/samsung-galaxy-s2-in-pakistan-teletec-stores). Android Pakistan (2011-07-23). Retrieved on 2011-11-28.
[21] "Samsung Likes Canada More Than US, First The Galaxy S II & Now This?" (http://www.gizmocrunch.com/android/8014-samsung-galaxy-s2-hercules-canada). *GizmoCrunch*. 18 August 2011. . Retrieved 20 August 2011.
[22] [LEAK] Samsung Galaxy S2 First Hits North America July 21st (http://www.gizmocrunch.com/android/7577-samsung-galaxy-s2-release-date). Gizmocrunch.com (2011-07-14). Retrieved on 2011-11-28.
[23] "Samsung Galaxy S II con Telcel $9,249.00 MXP" (http://www.poderpda.com/plataformas/android/samsung-galaxy-s-ii-llega-a-mexico-con-telcel/). *PoderPDA.com*. 28 June 2011. .
[24] Galaxy S II finally lands on American shores for Sprint, T-Mobile and AT&T (http://www.engadget.com/2011/08/30/galaxy-s-ii-finally-lands-on-american-shores-for-sprint-t-mobil/). Engadget. Retrieved on 2011-11-28.
[25] Samsung Galaxy S II: availability and pricing from operators « Cellular « MyBroadband Tech and IT News (http://mybroadband.co.za/news/cellular/26743-samsung-galaxy-s-ii-availability-and-pricing-from-operators.html). Mybroadband.co.za (2011-06-14). Retrieved on 2011-11-28.
[26] "Samsung's Exynos 4210 flexes 3D gaming muscle at GDC 2011 (video)" (http://www.engadget.com/2011/03/03/samsungs-exynos-4210-flexes-3d-gaming-muscle-at-gdc-2011-video/). *Engadget*. 3 March 2011. .
[27] "Samsung dubs its mobile processors Exynos, dual-core 4210 (formerly Orion) arriving next month" (http://www.engadget.com/2011/02/10/samsung-dubs-its-mobile-processors-exynos-dual-core-4210-forme/). *Engadget*. 10 February 2011. .
[28] "NEON – ARM" (http://mobile.arm.com/products/processors/technologies/neon.php?tab=Why+NEON). *ARM.com*. .
[29] "NVIDIA's Tegra 2 Take Two: More Architectural Details and Design Wins" (http://www.anandtech.com/show/4098/nvidias-tegra-2-take-two-more-architectural-details-and-design-wins/2). *Anandtech*. 5 January 2011. .
[30] Texas Instruments OMAP#OMAP 4
[31] UK ASA adjudication: Atrix is not world's most powerful as Samsung S II is more powerful (http://www.asa.org.uk/ASA-action/Adjudications/2011/8/Motorola-Mobility-UK-Ltd/SHP_ADJ_161483.aspx). Asa.org.uk. Retrieved on 2011-11-28.
[32] "TI OMAP-powered Samsung Galaxy S II I9100G pops up quietly" (http://www.gsmarena.com/ti_omappowered_samsung_galaxy_s_ii_i9100g_pops_up_quietly-news-3371.php). 11 November 2011. .
[33] "" (http://www.oled-info.com/super-amoled-plus). OLED-Info. . Retrieved 20110-12-04.
[34] Galaxy S II display reportedly has yellow spots, we test – GSMArena.com news (http://www.gsmarena.com/samsung_galaxy_s_ii_display_plagued_by_uneven_brightness_yellow_tint-news-2719.php). Gsmarena.com. Retrieved on 2011-11-28.
[35] "SAMSUNG GALAXY S II – REAL USER REVIEW PART 3: MEDIA" (http://blog.clove.co.uk/2011/05/31/samsung-galaxy-sii-real-user-review-part-3-media/). *Clove Technology's Blog*. 17 June 2011. .
[36] "Wolfson's innovative ultra low power audio hub selected for Samsung's latest Bada and Android smartphones" (http://www.wolfsonmicro.com/media_centre/item/wolfsons_innovative_ultra_low_power_audio_hub_selected_for_samsungs_latest_/). *Wolfson Microelectronics*. 17 June 2011. .
[37] "Silicon Summary in the Samsung Galaxy S II » Recent Teardowns » Chipworks" (http://www.chipworks.com/en/technical-competitive-analysis/resources/recent-teardowns/2011/07/silicon-summary-in-the-samsung-galaxy-s-ii/). *Chipworks*. 25 July 2011. .
[38] "Samsung GALAXY S II Features" (http://www.samsung.com/global/microsite/galaxys2/html/feature.html). *Samsung.com*. . Retrieved 21 April 2011.
[39] "Google unveils first Android NFC phone — but Nexus S is limited to tag reading only for now" (http://www.nearfieldcommunicationsworld.com/2010/12/07/35385/google-unveils-first-android-nfc-phone-but-nexus-s-is-limited-to-tag-reading-only-for-now/). *NFC World*. 17 June 2011. .
[40] "No NFC for UK Samsung Galaxy S2" (http://www.knowyourmobile.com/blog/867287/no_nfc_for_uk_samsung_galaxy_s2.html). *Know Your Mobile*. 28 April 2011. ,
[41] "Samsung officially confirms NFC-enabled Galaxy S II" (http://www.techradar.com/news/phone-and-communications/mobile-phones/samsung-officially-confirms-nfc-enabled-galaxy-s-ii-954965). *techradar.com*. 12 May 2011. ,
[42] Mobile High-Definition Link (MHL) (http://www.youtube.com/watch?v=ZtwTcnR0hqA). YouTube (2011-02-15). Retrieved on 2011-11-28.
[43] "Watch Samsung Galaxy S II connect to an HDTV through its MHL port [VIDEO (http://blog.gsmarena.com/watch-samsung-galaxy-s-ii-connect-to-hdtv-through-its-mhl-port-video/)"]. *GSMArena Blog*. 13 May 2011. .
[44] Samsung Galaxy S2 USB OTG demo (http://www.youtube.com/watch?v=giJXF5pIITc). YouTube. Retrieved on 2011-11-28.
[45] Broadcom's New Combo Chip Integrates 802.11n Wi-Fi, Bluetooth 3.0 + HS and FM to Bring New Multimedia Applications to Smartphones, Tablets and Other Mobile Devices (http://www.broadcom.com/press/release.php?id=s549642), Broadcom press release
[46] **(French)** "Des accessoires pour le Samsung Galaxy S2" (http://www.mysamsunggalaxys2.com). *Le Journal Du Geek*. 17 March 2011. .
[47] "Samsung Galaxy S™ II, Epic™ 4G Touch" (http://www.samsung.com/us/mobile/cell-phones/SPH-D710ZKASPR). *Samsung Mobile*. . Retrieved 1 September 2011.
[48] "Samsung Galaxy S™ II, available at AT&T" (http://www.samsung.com/us/mobile/cell-phones/SGH-I777ZKAATT). *Samsung Mobile*. . Retrieved 1 September 2011.
[49] "Samsung's Galaxy S II receives Android 2.3.6 update" (http://news.in.msn.com/technology/article.aspx?cp-documentid=5628795). MSN India. 28/11/2011. . Retrieved December 21, 2011.

[50] "Samsung Galaxy S II shows off motion-zoom option in TouchWiz 4.0 (video)" (http://www.engadget.com/2011/03/29/samsung-galaxy-s-ii-shows-off-gyro-zoom-option-in-touchwiz-4-0/). *Engadget*. 29 March 2011. .

[51] "Samsung Galaxy S2 Touchwiz 4.0 Special Features Demo [Exclusive (http://androidcommunity.com/samsung-galaxy-s2-touchwiz-4-0-special-features-demo-exclusive-20110325/)"]. *Android Community*. 25 March 2011. .

[52] "Samsung Galaxy S2: what you need to know" (http://mwc2011.techradar.com/2011/02/samsung-galaxy-s2-what-you-need-to-know/). *TechRadar*. 14 February 2011. .

[53] Samsung Kies Air Demo (Samsung Galaxy II) (http://www.youtube.com/watch?v=_qah8q631Jk). YouTube. Retrieved on 2011-11-28.

[54] Samsung Galaxy SII – Voice Translator Demo (http://www.youtube.com/watch?v=YjqwRPUMhBU). YouTube. Retrieved on 2011-11-28.

[55] "Big news for Android at Mobile World Congress" (http://blogs.sybase.com/afaria/2011/02/big-news-for-android-at-mobile-world-congress/). *Sybase Enterprise Mobility Blog*. 13 February 2011. .

[56] "AnyConnect for Samsung Devices – Android Market" (https://market.android.com/details?id=com.cisco.anyconnect.vpn.android). *Android Market (online)*. . Retrieved 20 August 2011.

[57] "Samsung I9100 Galaxy S II preview: Second encounter – Page 4" (http://www.gsmarena.com/samsung_i9100_galaxy_s_ii-review-588p4.php). *GSMArena.com*. 13 April 2011. .

[58] Samsung Galaxy S II Preview (http://www.youtube.com/watch?feature=player_detailpage&v=y4d-wWCf4CQ#t=416s). YouTube (2011-04-15). Retrieved on 2011-11-28.

[59] "HTC Sensation Review Vs Samsung Galaxy S II: Video playback shootout « Clove Technology's Blog" (http://blog.clove.co.uk/2011/07/06/htc-sensation-review-vs-samsung-galaxy-s-ii-video-playback-shootout/). *Clove*. 16 July 2011. .

[60] "CyanogenMod 7 for Samsung Galaxy S2 (II): Development Already Started!" (http://www.inspiredgeek.com/2011/06/08/cyanogenmod-7-for-samsung-galaxy-s2-ii-development-already-started/). *Inspired Geek*. 8 June 2011. .

[61] "CyanogenMod coming to the Galaxy S 2, thanks to Samsung" (http://www.androidcentral.com/cyanogenmod-coming-galaxy-sii-thanks-samsung). *Android Central*. 6 June 2011. .

[62] Savov, Vlad (28 April 2011). "Samsung Galaxy S II review" (http://www.engadget.com/2011/04/28/samsung-galaxy-s-ii-review/). *Engadget*. .

[63] Westaway, Luke (6 May 2011). "Samsung Galaxy S 2 review" (http://reviews.cnet.co.uk/mobile-phones/samsung-galaxy-s-2-review-50002442). *CNET UK*. . Retrieved 14 May 2011.

[64] Beavis, Gareth (26 April 2011). "Samsung Galaxy S2 Review" (http://www.techradar.com/reviews/phones/mobile-phones/samsung-galaxy-s2-930907/review?artc_pg=1). *TechRadar*. . Retrieved 14 May 2011.

[65] "Galaxy S II Web Browser Speed Test (Video)" (http://pocketnow.com/android/galaxy-s-ii-web-browser-speed-test-video). *Pocketnow.com*. 11 May 2011. .

[66] Davies, Chris (26 April 2011). "Samsung Galaxy S II Review" (http://www.slashgear.com/samsung-galaxy-s-ii-review-26148446/). *SlashGear*. . Retrieved 14 May 2011.

[67] "Samsung i9100 Galaxy S II review: Brightest star – Final words" (http://www.gsmarena.com/samsung_i9100_galaxy_s_ii-review-597p12.php). *GSMArena*. 13 May 2011. .

[68] "Galaxy S2 sales to consumers top 1 mln in S. Korea" (http://www.antaranews.com/en/news/72648/galaxy-s2-sales-to-consumers-top-1-mln-in-s-korea). *Antara News*. 13 June 2011. .

[69] "Galaxy S 2 Set a Record of 3 Million Global Sales in 55 Days" (http://www.flickr.com/photos/samsungtomorrow/5895933600/). *SamsungTomorrow Flickr*. 30 June 2011. .

[70] "Samsung Galaxy S2: Sales figures out of this world" (http://www.mobot.net/samsung-galaxy-s2-sales-figures-world-27840). *Mobot.net*. 8 August 2011. .

[71] "Samsung moves ten million Galaxy S II smartphones, pats itself on the back" (http://www.engadget.com/2011/09/25/samsung-moves-10-million-galaxy-s-iis-pats-itself-on-the-back/). *Engadget*. 25 September 2011. .

[72] 28 October 2011, Samsung overtakes Apple in smartphone sales (http://www.bbc.co.uk/news/business-15489523), BBC News

[73] "Samsung I9103 Galaxy R – Full phone specifications" (http://www.gsmarena.com/samsung_i9103_galaxy_r-3967.php). *GSMArena.com*. . Retrieved 22 October 2011.

[74] "Samsung takes the wraps off the Galaxy R smartphone" (http://www.gsmarena.com/samsung_takes_the_wraps_off_the_galaxy_r_smartphone-news-2947.php). *GSMArena.com*. 30 July 2011. . Retrieved 20 August 2011.

[75] "Samsung Galaxy W I8150 – Full phone specifications" (http://www.gsmarena.com/samsung_galaxy_w_i8150-4114.php). *GSMArena.com*. . Retrieved 22 October 2011.

[76] Samsung Galaxy W i8150 video preview ENG by HDblog (http://www.youtube.com/watch?v=RcYUQ_VEfLc). YouTube (2011-08-25). Retrieved on 2011-11-28.

[77] Bunker, Adam (8 September 2011). "Samsung Galaxy S2 Mini: April release date leaked by Three" (http://www.t3.com/news/samsung-galaxy-s2-mini-april-release-date-leaked-by-three). *T3 (t3.com)*. . Retrieved 22 October 2011.

[78] Hollister, Sean (20 May 2011). "Samsung Galaxy S II Mini leaks out for Three, plus Nokia X7, Flyer and PlayBook release dates in UK" (http://www.engadget.com/2011/03/20/samsung-galaxy-s-ii-mini-leaks-out-for-three-uk-plus-nokia-x7/). *Engadget*. . Retrieved 22 October 2011.

[79] Samsung Galaxy Player, Galaxy Tab 8.9-inch & 4G Phones Unveiled < Mobile Phones, Cell Phones, Samsung, Android, Mobile Phones, Tablets < PC World India News < PC World.in (http://www.pcworld.in/news/

samsung-galaxy-player-galaxy-tab-89-inch-4g-phones-unveiled-55472011). Pcworld.in (2011-09-27). Retrieved on 2011-11-28.
[80] Samsung Galaxy S II HD LTE and Galaxy S II LTE official (http://www.slashgear.com/samsung-galaxy-s-ii-hd-lte-and-galaxy-s-ii-lte-official-26182723/). SlashGear. Retrieved on 2011-11-28.
[81] "Samsung Galaxy S™ II, Epic™ 4G Touch fact sheet – Delivering rich content at blazing fast speeds" (http://newsroom.sprint.com/news/samsung-galaxy-s-ii-epic-4g-touch-fact-sheet.htm). *Sprint Newsroom*. 30 August 2011. . Retrieved 1 September 2011.
[82] "Sprint First U.S. Carrier to Debut Galaxy S II With Samsung Epic 4G Touch" (http://newsroom.sprint.com/news/sprint-first-us-carrier-to-debut-galaxy-s-ii-with-samsung-epic-4g-touch.htm). *Sprint Newsroom*. 30 August 2011. . Retrieved 1 September 2011.
[83] "Sprint Epic 4G Touch Lacking NFC Unlike AT&T and T-Mobile Models" (http://pocketnow.com/android/sprint-epic-4g-touch-lacking-nfc-unlike-att-and-t-mobile-models). *Pocketnow*. 3 September 2011. .
[84] "Samsung Galaxy S II Coming Soon to T-Mobile" (http://newsroom.t-mobile.com/articles/T-Mobile-confirms-Samsung-Galaxy-S-II). *T-Mobile*. 30 August 2011. . Retrieved 1 September 2011.
[85] "Samsung Galaxy S II logs confirm NFC support for AT&T, none for Sprint" (http://www.engadget.com/2011/08/31/samsung-galaxy-s-ii-logs-confirm-nfc-support-for-atandt-none-for/). 31 August 2011. .
[86] "HTC Amaze, Samsung Galaxy S II: T-Mobile's fastest 4G phones yet" (http://latimesblogs.latimes.com/technology/2011/09/t-mobiles-new-4g-phones-htc-amaze-and-samsung-galaxy-s-ii.html). *LA Times*. 26 September 2011. .
[87] http://www.samsung.com/us/mobile/cell-phones/SGH-I927ZKAATT
[88] "Confirmed: Rogers launching LTE Samsung Galaxy S II, plus LTE network in Toronto goes live September 28th." (http://mobilesyrup.com/2011/08/30/confirmed-rogers-launching-lte-samung-galaxy-s-ii-plus-lte-network-in-toronto-goes-live-september-28th/). *MobileSyrup*. 30 August 2011. .
[89] "4G Samsung Galaxy S II X – Bientôt disponible" (http://www.telusmobility.com/en/AB/samsung_galaxy_s_II_x/index.shtml). *Telus*. . Retrieved 27 September 2011.
[90] "Video: TELUS Samsung Galaxy S II X quick overview" (http://mobilesyrup.com/2011/09/22/video-telus-samsung-galaxy-s-ii-x-quick-overview/). *MobileSyrup*. 22 September 2011. .
[91] Merrett, Andy (1 June 2011). "Samsung Galaxy S3 to launch in early 2012, 4G tablet coming sooner" (http://crave.cnet.co.uk/mobiles/samsung-galaxy-s3-to-launch-in-early-2012-4g-tablet-coming-sooner-50003954/). *crave cnet 2012*. CNET UK. . Retrieved 4 August 2011.
[92] Kennemer, Quentyn (30 May 2011). "Samsung Prepping Samsung Galaxy S III for First Half of 2012" (http://phandroid.com/2011/05/30/samsung-prepping-samsung-galaxy-s-iii-for-first-half-of-2012/). *phandroid gal 2012*. phandroid.com. . Retrieved 4 August 2011.

External links

- Samsung Galaxy S2 (http://technucleus.com/samsung-galaxy-s2.html) - Perfect Viewing Experience
- Samsung Galaxy S II Homepage (http://www.samsung.com/global/microsite/galaxys2/html/index.html)
- An Ideal Phone (http://galaxys2smarttouchphone.vanquishx.com) - Packed with Features

Samsung_Galaxy_Tab

The initial release of the Galaxy Tab.

Developer	Samsung
Manufacturer	Samsung Electronics
Type	Tablet/Media player/PC
Release date	Varies by region
Operating system	Android 2.2.1 with TouchWiz UI. Upgradeable to Android 2.3.3/2.3.4/2.3.5/2.3.7 in some countries (not compatible with 3.0).
Power	4000 mAh battery
CPU	Samsung Exynos 3110 (code-named "Hummingbird") ARM Cortex A8; 1.0 GHz
Storage capacity	Flash memory 2 GB (CDMA), 16 GB or 32 GB models and microSD slot
Memory	512 MB
Display	1024 × 600 px (aspect ratio 16:10), 7.0 in (18 cm) diagonal, appr. 21 in^2 (140 cm^2) at 170 PPI
Graphics	PowerVR SGX 540
Input	Multi-touch screen
Camera	3.2 MP AF camera with LED flash, 1.3 MP front-facing (for video calls)
Connectivity	(GSM/ GPRS/ EDGE) : 850 / 900 / 1800 / 1900 MHz (HSUPA 5.76Mbps, HSDPA 7.2Mbps) : 900 / 1900 / 2100 MHz, or 1700 / 1900 MHz CDMA 800 / 1900 MHz EVDO Rev A[1] Wi-Fi 802.11b/g/n, Bluetooth 3.0, DLNA Connection Port
Dimensions	190.09 mm (7.484 in) *(h)* 120.45 mm (4.742 in) *(w)* 11.98 mm (0.472 in) *(d)*
Weight	380 g (13 oz)
Related articles	Samsung Galaxy S Samsung Galaxy Tab 10.1 Samsung Galaxy Tab 7.7
Website	Samsung GALAXY Tab [2]

The **Samsung Galaxy Tab** is an Android-based tablet computer produced by Samsung[3] introduced[4] on 2 September 2010 at the IFA in Berlin.

The Galaxy Tab has a 7-inch (180 mm) TFT-LCD touchscreen, Wi-Fi capability, a 1.0 GHz ARM Cortex-A8 Samsung Exynos 3110 (code-named "Hummingbird") processor, the Swype input system,[5] a 3.2 MP rear-facing camera and a 1.3 MP front-facing camera for video calls. It runs the Android 2.2 (Froyo) operating system,[6] and supports telephone functionality as speaker phone, via provided wired ear piece or Bluetooth earpieces (except models sold in the US). It can download videoconferencing apps such as Tango as alternative to telephone functionality.[7]

Hardware

The tablet is enclosed in a plastic frame[8] that makes it lighter than other metal-bodied tablets, weighing 380 g (0.84 lb).[9]

The GT-P1000 model carries a 7" Super TFT instead of the AMOLED which is used by Samsung in its Galaxy S phones.[10] The screen has a 1024×600 resolution With mDNIE (Mobile Digital Natural Images Engine). Internal flash storage of 2 GB (North America CDMA models), 16 GB or 32 GB can be supplemented with a microSD flash card with up to 32 GB. CPU is a Exynos 3110 Applications Processor (Also known as Hummingbird) features 1.0 GHz ARM architecture Cortex A8 application and has 512 MB of RAM paired with a PowerVR SGX540 graphics processor.

The WIFI only model has a different graphics chip that doesn't support TV out and Samsung has not provided (or promised to provide) an update to gingerbread.

The tablet has two cameras: a 3.2 rear MP camera with a LED flash and a 1.3 MP front camera for video calling (the Verizon model has a 3 megapixel rear camera).[11] The front camera has auto focus capability. The camera also has auto image stitching, combining 8 pictures. Modes include single shot, continuous, panorama, and self-shot. It can automatically trigger on detecting that the subject smiles. Autogeotagging uses the internal GPS receiver.

The tablet has GPS, 802.11n Wi-Fi, Bluetooth 3.0, and handsfree/bluetooth/headphone telephony. Cellular protocols include GSM CDMA, HSPA (HSUPA).

It also has a 30-pin docking and charging connector very similar to the standard PDMI connector (a non-proprietary alternative to Apple's docking connector). It appears so similar to the PDMI connector that it is widely mistaken for it, but it is non-standard and all accessories, including charging cables, are incompatible with other equipment and only available from Samsung.

Samsung says that its 4000 mAh battery will give it 7 hours of video playback or 10 hours of talk time.[12] [13]

The GSM variants of the Galaxy Tab have an externally-accessible SIM card slot. If the SIM card is removed while the system is on, the system automatically reboots. The AT&T and T-Mobile variants of the Galaxy Tab ship with a micro SIM in a micro SIM adapter. For non-US Galaxy Tabs, with phone function, this slot can also accommodate a 3G data-only SIM card if the user does not need telephone functionality.

The Samsung Galaxy Tab also has an optional RCA plug connector through which the screen image is shown on a TV or other display (the Tab's own screen cannot be turned off; it can be dimmed with a backlight dimming app).[14]

Software

This tablet comes with a version of the Android 2.2 operating system with some custom skins and applications.[15] Most Android 2.2 apps developed using Google's guidelines for Android should scale properly when displayed on larger-screen devices such as the Samsung Galaxy Tab, according to Samsung.[16] Adobe Flash 10.1, DivX,[17] MPEG-4, WMV and Xvid, H.263, H.264[18] support has also been announced. The Tab uses Atmel's maXTouch multi touch capacitive touchscreen,[19] [20] and supports multi-tasking.[21] [22]

The Tab supports calendar, email and instant messaging applications. It has a launcher for e-reading applications which starts PressDisplay when reading newspapers, Kobo when reading e-books, and Zinio when reading magazines.[23] For viewing and editing Microsoft Office documents, the Galaxy Tab also comes bundled with the Android version of ThinkFree Office Mobile.[24]

The Tab can also provide tethering, acting as a Wi-Fi hotspot for up to 5 devices. Stored addresses can be displayed in Google maps with one click. It also allows linkage with a contact's Facebook profile if the phone address is linked with the contact's Facebook address.

Several HD video content multimedia formats, including DivX, XviD, MPEG4, H.263, H.264, are supported.[25] It can play video content either stored on the device itself or streamed from YouTube, and can output 720p video to a TV either as composite video or via HDMI when using the optional dock.[26]

Text can be entered using Swype, by tracing a path over letters on a virtual keyboard, and standard XT9 predictive typing is also supported.[27]

TouchWiz Samsung Galaxy Tabs allow the screen display to be saved.[28] Honeycomb tablets without TouchWiz, including the 10.1 in Galaxy Tab, must either be rooted or have the SDK installed in order to capture screenshots [29]

Upgrades

In May 2011 it was reported that Android Gingerbread 2.3.3 was being made available in Italy, with other regions expected to follow.[30]

Android Gingerbread 2.3.4 is available for the T-Mobile version.

Early impressions and reviews

Some published early impressions were favourable,[16] and the Tab was considered a serious rival to the iPad.[31] [32] "Rough edges" which should improve with later software updates were commented on.[33]

Other early reviews were more critical; one reviewer commented "The Tab ... [merges] the worst of a tablet and the worst of a phone."[34]

Release

Africa

- South Africa started selling the Tablet from November 2010[35]
- Nigeria | 5 November 2010, Etisalat Nigeria became the first cellular network in Africa to launch the Galaxy Tab.[36]
- Angola 1 December 2010, Movicel used the Samsung Galaxy Tab 7" as the launch Tablet for its December launched GSM 900 Mhz Network conversion from a CDMA 800 MHz.
- Ghana | 6 December 2010, Vodafone Ghana launched the Galaxy Tab.

Asia

The Samsung Galaxy Tab has been released in Indonesia and Thailand without contract and in Malaysia under contract by Maxis. While in the Philippines under contract by Smart and is also available without contract for GSM version. Under contract with NTT Docomo in Japan. In Singapore, it was released exclusively with Singtel on 13 November 2010.

South Korea

The Samsung Galaxy Tab was released South Korea on 3 November 2010, delayed from the original release date of 14 October.

Japan

The Samsung Galaxy Tab was released on 26 November 2010.

Pakistan

The Samsung Galaxy Tab will be released on 21 December 2010. It is available without a contract from Mobilink as well as in the open market.

India

Samsung launched the Galaxy Tab in India on 10th of November 2010.

Australia

The Samsung Galaxy Tab was released on 8 November 2010. It is available without a contract from several major national retailers, and under contract from Telstra, Optus and other carriers.[37]

Brazil

The Samsung Galaxy Tab was released in Brazil on 26 November 2010. Brazilian version seems to be faster; the processor runs @ 1.2 GHz and it has support for Analog TV and Digital SBTVD.

Europe

The Samsung Galaxy Tab was released in Germany and Poland on 11 October 2010. The UK version of the Tab was released on 1 November 2010. In Spain and most other European countries it was released later in 2010.

Middle East

The Samsung Galaxy Tab has been officially released in the Middle east in Dubai's GITEX Shopper the largest gathering of the region's leading ICT retailers and suppliers.

United States

The Galaxy tab has been released in the US from T-Mobile, Sprint, AT&T, U.S. Cellular and Verizon.

T-Mobile started offering the SGH-T849 Galaxy Tab on 10 November 2010[38]

Verizon started offering the SCH-I800 Galaxy Tab on 11 November 2010. Verizon's version of the Galaxy Tab has a textured black back, as opposed to the standard plastic rear of the T-Mobile, Sprint, and AT&T versions. The Verizon and Sprint versions disable the Human Interface Device (HID) bluetooth features so it will not work with bluetooth keyboards, mice, etc. However, updates did correct this issue for both carriers.

Sprint started offering the Galaxy Tab on 14 November 2010.

AT&T began offering the Tab on 21 November 2010, with no contract requirement.[39]

The FCC approved a Wi-Fi only version of the Galaxy Tab.[40] Some Best Buy advertisements appear to have been prematurely leaked about the Wi-Fi only version of the Galaxy Tab.[41]

Wi-Fi-only model was released on 11 April 2011.[42]

Sales

A week after its release, Samsung announced that they had sold 600,000 units.[43] On the 4 December, it was reported that the 1 million mark was reached, two months after launch.[44] In January 2011 Samsung announced they had shipped 2 million units to stores.

Successor models

During the 2011 International Consumer Electronics Show (CES), Verizon Wireless and Samsung Telecommunications America (Samsung Mobile) announced that a new 4G LTE-Enabled Samsung Galaxy Tab features access to Verizon Wireless' 4G LTE Mobile Broadband Network and a 5 megapixel rear-facing Camera will be available.[45] [46]

At the Mobile World Congress event in 2011 Barcelona Samsung showed a new Galaxy Tab model. It features a bigger 10.1 inch HD display with a Dual-Core NVIDIA Tegra 2 processor, running Google's Android Honeycomb operating system.[47] It was set for a US release in March 2011 and a European release in April. However, after the iPad 2 release, some specifications were described as "*inadequate*" by Lee Don-Joo,[48] executive vice president of Samsung's mobile division, pointing to a possible model review or rethink of their market strategy.

This would lead to the introduction of a newer, slimmer 10.1" model at the Samsung Unpacked Event during CTIA Wireless Convention in March 2011, together with a 8.9 inch model, pushing the release date further to 8 June for the US release and "early summer" for the latter model. Although there was no information about a delay of the European release date, it was announced that the previous design, seen at the Mobile World Congress, would be sold relabelled as "Samsung Galaxy Tab 10.1v".[49] [50]

During IFA 2011 in Berlin, Samsung announced the new Galaxy Tab 7.7, sporting a dual-core 1.4 GHz processor, 1GB of memory, support for 32GB MicroSD cards (possibly only on some models), a 5,100mAh battery, a front-facing 2-megapixel camera and a rear-facing 3-megapixel camera with flash. There will be 16GB, 32GB and 64GB models available and each will support 802.11 a/b/g/n standards at both 2.4 GHz and 5 GHz. The unit measures 196.7 x 133 x 7.89 mm (7.75 x 5.24 x 0.31 inches) and weighs 335g (11.8 oz) making it much more single-hand friendly than larger models from Samsung and other manufacturers.[51]

The final successor to the original Galaxy Tab is the Samsung Galaxy Tab 7.0 Plus. This model features the same 7" screen with a bit smaller resolution of 1,024 x 600 and uses PLS panel. Also it now comes with Android 3.2 Honeycomb pre-installed as well as including newer, more powerful hardware. The new model includes a 1.2GHz dual-core processor, 16GB or 32GB of user accessible memory (RAM), and a 2-megapixel front- and 3-megapixel rear-facing camera. The device is much smaller than the original, measuring 193.5 x 122.4 x 9.9 mm and weighing just 345 grams.

See also

- List of Android devices
- Tablet PC

References

[1] "Galaxy Tab" (http://www.samsung.com/us/mobile/galaxy-tab). . Retrieved 24 Dec 2010.

[2] http://www.samsung.com/global/microsite/galaxytab/2010/

[3] "Galaxy Tab unveiled as Samsung's first tablet computer" (http://www.bbc.co.uk/news/technology-11163687). BBC News. 2 September 2010. . Retrieved 4 Dec 2010.

[4] Tim Gideon (24 August 2010). "Samsung Galaxy Tablet Coming in September" (http://www.pcmag.com/article2/0,2817,2368214,00.asp). *PC Magazine*. Ziff Davis. . Retrieved 4 Dec 2010.

[5] Ben Woods (25 August 2010). "Samsung teases Galaxy Tab tablet device features" (http://www.zdnet.co.uk/news/mobile-devices/2010/08/25/samsung-teases-galaxy-tab-tablet-device-features-40089916). *ZDNet UK*. Ziff Davis. . Retrieved 4 Dec 2010.

[6] Arthur Naresh. "Samsung Galaxy Tablet – Features and Specifications" (https://acrobat.com/#d=2azFVkcVDidx*FKB7n4RAA). . Retrieved 4 Dec 2010.

[7] 4 Android tablets vie for your attention (http://www.computerworld.com/s/article/9201721/4_Android_tablets_vie_for_your_attention?taxonomyId=15&pageNumber=3). *Computerworld* (22 December 2010). Retrieved on 3 July 2011.

[8] Chris Davies (31 October 2010). "Samsung Galaxy Tab Review" (http://www.slashgear.com/samsung-galaxy-tab-review-31111323/). *SlashGear*. R3 Media LLC. . Retrieved 4 Dec 2010.

[9] Graeme Wearden (2 September 2010). "Samsung Galaxy Tab revealed at IFA" (http://www.guardian.co.uk/technology/2010/sep/02/samsung-galaxy-tab-revealed-ifa). *The Guardian*. UK. . Retrieved 4 Dec 2010.

[10] Mikael Ricknäs (2 September 2010). "Samsung launches Galaxy Tab" (http://www.techworld.com.au/article/359327/samsung_launches_galaxy_tab). *TechWorld*. IDG. . Retrieved 4 Dec 2010.

[11] "Samsung Galaxy Tab™" (http://www.verizonwireless.com/b2c/store/controller?item=phoneFirst&action=viewPhoneDetail&selectedPhoneId=5565&deviceCategoryId=12). Verizon. .

[12] "Samsung Galaxy Tab" (http://www.samsung.com/hk_en/consumer/mobile/mobile-phones/mobile-tablet/GT-P1000CWATGY/index.idx?pagetype=prd_detail&tab=specification). *Samsung*. . Retrieved 4 Dec 2010.

[13] "Samsung Galaxy Tab" (http://samsungtabletpc.org). .

[14] Samsung Galaxy Tab RCA Video Cable Review (http://www.thegalaxytab.com/2011/01/01/samsung-galaxy-tab-rca-video-cable-review/). Thegalaxytab.com (1 January 2011). Retrieved on 3 July 2011.

[15] "Samsung Launches Galaxy Tab" (http://www.pcworld.com/article/204729/samsung_launches_galaxy_tab.html). PC World. 3 September 2010. . Retrieved 5 September 2010.

[16] "Critics' Choice: First Hands-on Assessments of Samsung Galaxy Tab" (http://www.enterprisemobiletoday.com/features/article.php/3904596/Critics-Choice-First-Hands-on-Assessments-of-Samsung-Galaxy-Tab.htm). EnterpriseMobileToday. 21 September 2010. . Retrieved 23 March 2011.

[17] "Samsung Galaxy Tab: first impressions" (http://blogs.ft.com/techblog/2010/09/samsung-galaxy-tab-first-impressions/). *Financial Times*. . Retrieved 2 September 2010.

[18] Samsung Galaxy Tab"http://www.vgaze.com/2010/09/samsung-galaxy-tab.html"

[19] Atmel confirms the Samsung Galaxy Tab uses its maXTouch touchscreen controller (http://www.engadget.com/2010/09/27/atmel-confirms-the-samsung-galaxy-tab-uses-its-maxtouch-touchscr/). Engadget.com (27 September 2010). Retrieved on 3 July 2011.

[20] Latest Android 2.2 Froyo Multi touch capacitive Android tablet (A-OK PAD) (http://beon.en.alibaba.com/product/343028135-210181541/Latest_Android_2_2_Froyo_Multi_touch_capacitive_Android_tablet_A_OK_PAD_.html). Beon.en.alibaba.com (1 January 2010). Retrieved on 3 July 2011.

[21] Samsung Galaxy Tab for launch in US, Multitasking Maestro can beat iPad with new iOS (http://ic-technews.com/mobile-gadgets/1423-samsung-galaxy-tab-for-launch-in-us-multitasking-maestro-can-beat-ipad-with-new-ios). Ic-technews.com (19 September 2010). Retrieved on 3 July 2011.

[22] Samsung Galaxy Tab Deals: Get Awesome Multitasking Device At Reasonable Rate! (http://www.articlesnatch.com/Article/Samsung-Galaxy-Tab-Deals--Get-Awesome-Multitasking-Device-At-Reasonable-Rate-/1950922). Articlesnatch.com. Retrieved on 3 July 2011.

[23] "Samsung Galaxy Tab preview" (http://www.engadget.com/2010/09/02/samsung-galaxy-tab-preview/). Engadget. 2 September 2010. .

[24] "Galaxy Tab – Overview – Samsung Mobile Singapore" (http://www.samsungmobile.com.sg/mobile-phones/galaxy-tab). Samsung Asia Pte Ltd. . Retrieved 3 December 2010.

[25] Do More on the Go with the Galaxy Tab (http://www.samsung.com/us/article/do-more-on-the-go-with-the-galaxy-tab). Samsung.com. Retrieved on 3 July 2011.

[26] Samsung Galaxy Tab HDMI Multimedia Dock (http://www.cellfreeks.com/productdetail.asp?productid=447). Cellfreeks.com. Retrieved on 3 July 2011.

[27] Samsung Galaxy Tab: An Android contender | 3 September 2010 (22:59) | Technology | By: BlackCode (http://computerszine.com/samsung-galaxy-tab-an-android-contender). Computerszine.com (3 September 2010). Retrieved on 3 July 2011.

[28] Hands-on with the new Android Market on a Samsung Galaxy Tab, By Matthew Miller (http://www.zdnet.com/blog/mobile-gadgeteer/hands-on-with-the-new-android-market-on-a-galaxy-tab/4203?tag=mantle_skin;content). Zdnet.com. Retrieved on 3 July 2011.

[29] http://howto.cnet.com/8301-11310_39-20080737-285/how-to-take-screenshots-on-your-samsung-galaxy-tab-10.1/?tag=mncol;mlt_related

[30] "Report: Samsung Galaxy Tab 2.3.3 Gingerbread update begins international rollout" (http://www.engadget.com/2011/05/12/samsung-galaxy-tab-2-3-3-gingerbread-update-begins-international/). Engadget. 12 May 2011. .

[31] "Samsung Galaxy Tab: First Tablet to Make iPad Sweat" (http://blog.laptopmag.com/samsung-galaxy-tab-first-tablet-to-make-ipad-sweat). Laptop. 2 September 2010. . Retrieved 23 March 2011.

[32] "Apple bans Samsung Galaxy – Is Samsung Winning the War Against Apple?" (http://techians.com/articles/apple-bans-samsung-galaxy-is-samsung-winning-the-war-against-apple/). Techians Blog. 6 September 2011. .

[33] Jason Perlow (11 November 2010). "Samsung Galaxy Tab: iPad Rival or Handheld Computer?" (http://www.zdnet.com/blog/perlow/samsung-galaxy-tab-ipad-rival-or-handheld-computer/14579). . Retrieved 12 November 2010.

[34] Matt Buchanan (10 November 2010). "Samsung Galaxy Tab Review: A Pocketable Train Wreck" (http://gizmodo.com/#!5686161/samsung-galaxy-tab-review-a-pocketable-train-wreck). . Retrieved 11 February 2011.
[35] Galaxy Tablet in South Africa (http://www.bcnn5.com/2010/11/samsung-galaxy-tablet-now-available-in-south-africa.html). Bcnn5.com (28 November 2010). Retrieved on 3 July 2011.
[36] "Experience the Samsung Galaxy Tab" (http://www.etisalat.com.ng/device_galaxy.php). etisalat.com.ng. 22 November 2010. .
[37] LeMay, Renai (8 August 2010). "Samsung Galaxy Tab Australian pricing" (http://www.cio.com.au/article/367185/samsung_galaxy_tab_australian_pricing/). .
[38] "T-Mobile gets first dibs on Galaxy Tab" (http://gizmodo.com/5674534/t+mobile-gets-first-dibs-on-galaxy-tab-november-10th-for-400). gizmodo.com. 27 October 2010. .
[39] "Samsung Galaxy Tab now available through AT&T" (http://www.csmonitor.com/Innovation/Horizons/2010/1122/Samsung-Galaxy-Tab-now-available-through-AT-T). CSMonitor.com. 22 November 2010. . Retrieved 23 March 2011.
[40] "Samsung GT-P1010 WiFi-only Galaxy Tab clears FCC" (http://www.slashgear.com/samsung-gt-p1010-wifi-only-galaxy-tab-clears-fcc-29111039/). SlashGear. . Retrieved 23 March 2011.
[41] "WiFi-only Samsung Galaxy Tab listed at Best Buy" (http://www.slashgear.com/wifi-only-samsung-galaxy-tab-listed-at-best-buy-25110008/). SlashGear. . Retrieved 23 March 2011.
[42] "Wifi-only Galaxy Tab officially available from Samsung" (http://www.androidcentral.com/wifi-only-galaxy-tab-officially-available-samsung). AndroidCentral. .
[43] "Samsung: Galaxy Tab sales "robust," head 1 million sales before year end" (http://www.zdnet.com/blog/btl/samsung-galaxy-tab-sales-robust-headed-for-1-million-sales-before-year-end/42036). zdnet.com. 29 November 2010. .
[44] "Galaxy Tab Sells 1 Million Units" (http://www.pcworld.com/article/212493/galaxy_tab_sells_1_million_units.html). PCWorld. 4 December 2010. .
[45] VERIZON WIRELESS AND SAMSUNG MOBILE ANNOUNCES FIRST 4G LTE-ENABLED SAMSUNG GALAXY TAB (http://www.mobile88.com/news/read.asp?file=/2011/1/7/20110106215414&phone=Samsung_Galaxy_Tab-Verizon-LTE). Mobile88.com (7 January 2011). Retrieved on 3 July 2011.
[46] The 4G Galaxy Tab got a silent upgrade. Did you notice? (http://thenextweb.com/mobile/2011/01/06/the-4g-galaxy-tab-got-a-silent-upgrade-did-you-notice/). Thenextweb.com (30 November 2010). Retrieved on 3 July 2011.
[47] "Samsung announces slimmed down Galaxy S II smartphone and upsized Galaxy Tab 10.1 tablet" (http://www.gizmag.com/samsung-announces-slimmed-down-galaxy-s-ii-smartphone-and-upsized-galaxy-tab-101-tablet/17865/picture/130201/). Gizmag.com. . Retrieved 23 March 2011.
[48] Lee Youkyoung, "Samsung sees iPad 2's thinness, price as challenges" (http://english.yonhapnews.co.kr/techscience/2011/03/04/9/0601000000AEN20110304009300320F.HTML), *Yonhap News Agency*, 4 March 2011
[49] Chris Davies , "Samsung Galaxy Tab 10.1V heads to Europe" (http://www.slashgear.com/samsung-galaxy-tab-10-1v-heads-to-europe-22141685/), *Slashgear*, 22 March 2011
[50] "Samsung Tablet" (http://samsungtabletpc.org/samsung-tablet-pcâdifference-between-samsung-galaxy-tab-10-1-10-1v/)
[51] "Samsung Galaxy Tab 7.7" (http://global.samsungtomorrow.com/?p=5292)

External links

- Samsung Galaxy Tab official page (http://galaxytab.samsungmobile.com/2010/index.html)
- Samsung Galaxy Tab Fans Page (http://mygalaxytab.net)

ref>[http://www.mobile88.com/news/read.asp?file=/2011/1/7/20110106215414

Mobile_phone_form_factors

Mobile phones are designed with one of a variety of different form factors, depending on its functionality or aesthetic considerations.

Rigid form factors

Bar

A **bar** (**slab**, **block**, or, commonly in the United States, **candybar**) phone takes the shape of a cuboid.[1] [2] It is named because of its resemblance to a candy bar in size and shape. This form factor is widely used by a variety of manufacturers, such as Nokia and Sony Ericsson. Bar-type mobile phones normally have the screen and keypad all on one face. The Samsung SPH-M620 is a unique take on the bar form factor, offering different devices on either side of the bar: a phone on one side, and a digital audio player on the other.

The Nokia E51, a typical "candybar" phone.

Slate

A **slate** phone is a subset of the bar form that, like a slate computer, has minimal buttons, instead relying upon a touchscreen and virtual QWERTY keyboard. Well-known slate smartphone manufacturers are LG Electronics, Apple, HTC, Samsung Mobile, and Nokia.

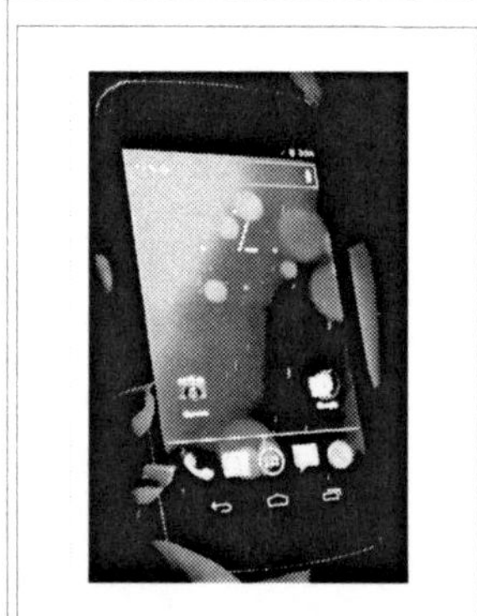

Galaxy Nexus, a slate Android smartphone

iPhone 4S, the Apple smartphone with iOS

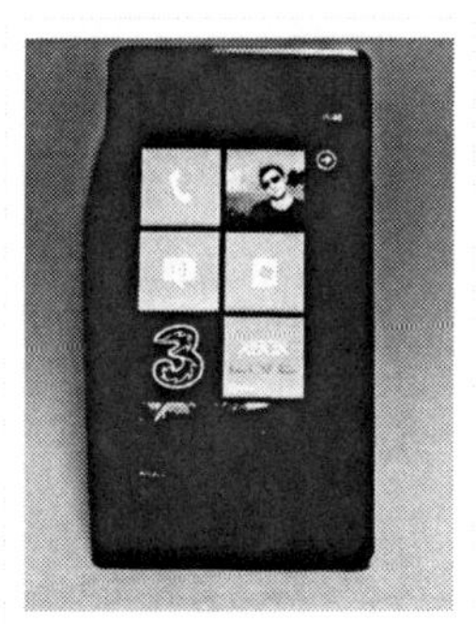

Samsung Omnia 7, a Windows Phone 7 slate smartphone

Moving form factors

Flip

A **flip** (or **clamshell**) phone consists of two or more sections that are connected by hinges, allowing the phone to fold or "flip" in order to become more compact. When flipped open, the phone's speaker and microphone are placed closer to the operator's ear and mouth, improving usability. When flipped shut, the phone can become much smaller and more portable than when it is opened for use.

The Sony Ericsson W980, a flip phone

Motorola was once owner of a trademark for the term "flip phone",[3] but the term "flip phone" has become genericized, and used more frequently than "clamshell" in colloquial speech. Motorola was the manufacturer of the famed StarTAC flip phone.

In 2010, Motorola introduced a different take on the flip phone with its Backflip smartphone. When closed, one side is the screen and the other is a physical QWERTY keyboard. The hinge is on a long edge of the phone instead of a short edge, and when flipped out, the screen is above the keyboard.

Slider

BlackBerry Torch 9800, a tall slider

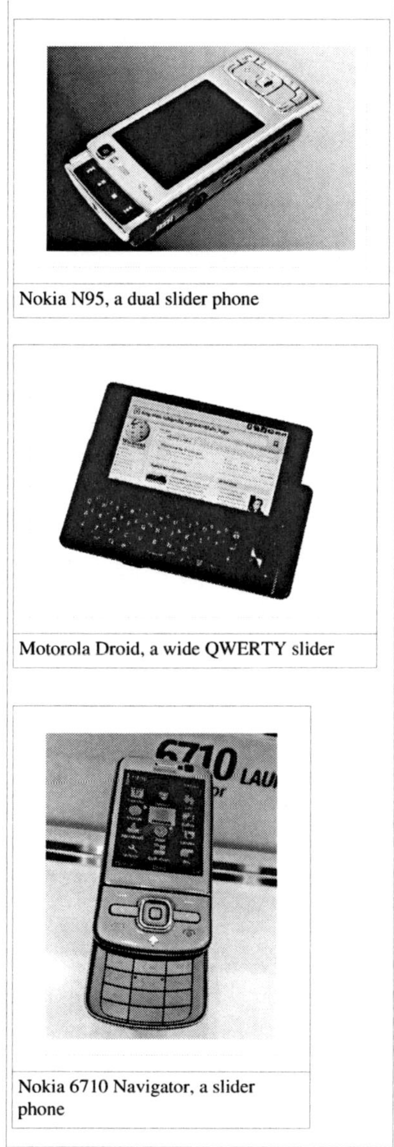

Nokia N95, a dual slider phone

Motorola Droid, a wide QWERTY slider

Nokia 6710 Navigator, a slider phone

A **slider** or **slide** phone is composed of usually two, but sometimes more, sections that slide past each other on rails. Most slider phones have a display segment which houses the speaker used for calls and the phone's screen, while another segment contains the keypad or keyboard and slides out for use. The goal of using a sliding form factor is to allow the operator to take advantage of full physical keyboards or keypads, without sacrificing portability, by "retracting" them into the phone when these are not in use.

The Siemens SL10 was one of the first sliding mobile phones in 1999. Some phones have an automatic slider built in which deploys the keypad. Many phones will "pop out" their keypad segments as soon as the user begins to slide the phone apart. Unique models are the 2-way slider (sliding up or down provides distinct functions) such as the Nokia N85 or Nokia N95.

A version of the slider form factor, the side slider or QWERTY slider, uses vertical access of the keyboard on the bottom segment. The side slider form factor is primarily used to facilitate faster access to the keyboard with both thumbs. The Danger Hiptop and the Sony Mylo are two primary examples.

Swivel

A **swivel** phone is composed of multiple (usually two) segments, which swivel past each other about a central axle. Use of the swiveling form factor has similar goals to that of the slider, but this form factor is less widely used.

The Motorola Flipout, a swivel phone

Mixed

Some phone models use both a swivel and a flip axis, like the Nokia N90.

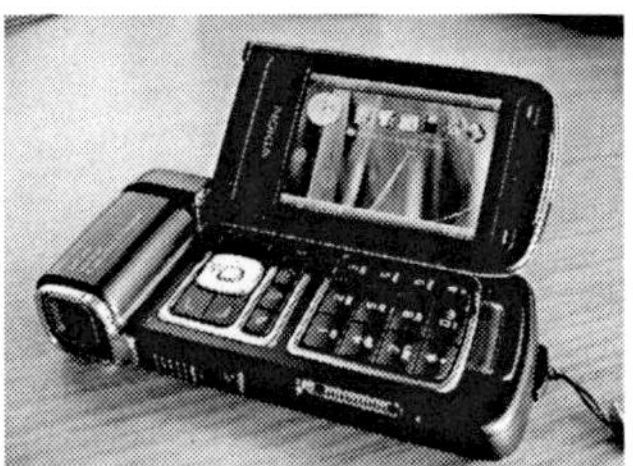

Nokia N93, a mixed phone

Brick

"Brick" is a slang term often used to refer to large, outdated bar-type phones, though it can be applied to older flip, slider and swivel phones as well.[4] [5] [6] Large, bulky phones such as the Motorola DynaTAC have been displaced by their newer, smaller counterparts, which offer greater portability thanks to smaller antennas and slimmer battery packs.

References

[1] Samsung showcases T509 Slim Bar Phone (http://www.mobiletechnews.com/info/2006/04/06/114802.html); [2006-04-06]; retrieved on [2008-05-18]

[2] 8310 Phone Review (http://www.conita.com/Mobile-Phones/Nokia/48.html); [2007-06-08]; retrieved on [2008-05-18]

[3] US trademark #2157939, cancelled February 26, 2005

[4] Associated Press (2005-04-11). "First cell phone a true 'brick'" (http://www.msnbc.msn.com/id/7432915). MSNBC. . Retrieved 2008-05-18.

[5] Olson, Darrin (2007-02-19). "80's Brick Cell Phone" (http://www.slipperybrick.com/2007/02/80s-brick-cell-phone). *SlipperyBrick.com*. Pragmatic Labs. . Retrieved 2008-05-18.

[6] "DynaTAC 8000X - the World's First Mobile Phone" (http://www.bbc.co.uk/dna/h2g2/A1082521). *h2g2*. BBC. 2003-07-03. . Retrieved 2008-05-18.

Subscriber_identity_module

A **subscriber identity module** or **subscriber identification module** (**SIM**) is an integrated circuit that securely stores the International Mobile Subscriber Identity (IMSI) and the related key used to identify and authenticate subscribers on mobile telephony devices (such as mobile phones and computers).

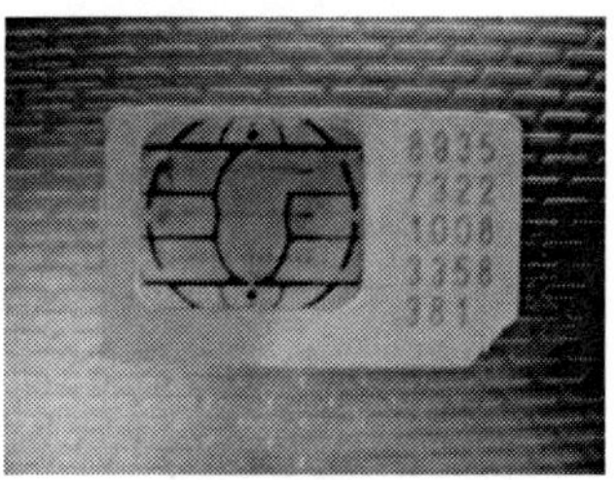

A typical SIM card

A SIM is embedded into a removable *SIM card*, which can be transferred between different mobile devices. SIM cards were first made the same size as a credit card (85.60 mm × 53.98 mm × 0.76 mm). The development of physically-smaller mobile devices prompted the development of a smaller SIM card, the mini-SIM card. Mini-SIM cards have the same thickness as full-size cards, but their length and width are reduced to 25 mm × 15 mm.

A SIM card contains its unique serial number (ICCID), internationally unique number of the mobile user (IMSI), security authentication and ciphering information, temporary information related to the local network, a list of the services the user has access to and two passwords: a personal identification number (PIN) for ordinary use and a personal unblocking code (PUK) for PIN unlocking.

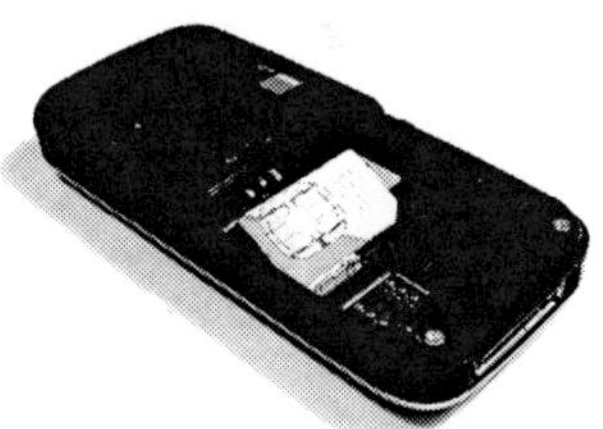

A mini-SIM card next to its electrical contacts in a Nokia 6233

History

The first SIM card was made in 1991 by Munich smart-card maker Giesecke & Devrient, who sold the first 300 SIM cards to the Finnish wireless network operator Radiolinja.[1]

The specification that standardized the micro-SIM form factor continues to evolve. Some features introduced recently include:

- A micro-SIM form factor (i.e. size, shape, and having essential physical details)
- Allowance for multiple simultaneous applications accessing the card through logical channels
- Mutual authentication as a way to eliminate carrier spoofing by allowing the SIM card to authenticate the cell tower to which it is connecting
- PIN protection with hierarchical PIN management with a universal PIN, an application PIN and a local PIN
- Expanded phonebook storage on the SIM card with entries for email, second name, and groups[2]

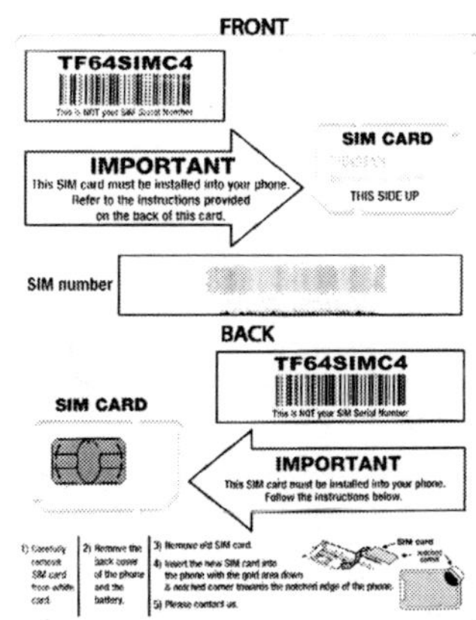

A TracFone Wireless SIM card has no distinctive carrier markings and is only marked as a "SIM CARD"

Design

There are three operating voltages for SIM cards: 5 V, 3 V and 1.8 V (ISO/IEC 7816-3 classes A, B and C, respectively). The operating voltage of the majority of SIM cards launched before 1998 was 5 V. SIM cards produced subsequently are compatible with 3 V and 5 V or with 1.8 V and 3 V.

SIM chip structure and packaging

Dual SIM phones are now made by some mobile phone manufacturers, which save the user from carrying around a separate phone for every number. There are two types: the first allows one to switch between the SIMs, whilst the second allows both SIMs to be active simultaneously.

SIM operating systems come in two main types: native and Java Card. Native SIMs are based on proprietary, vendor-specific software, whereas the Java Card SIMs are based on standards, particularly Java Card, which is a subset of the Java programming language specifically targeted at embedded devices.[3] Java Card allows the SIM to contain programs that are hardware independent and interoperable.

Data

SIM cards store network-specific information used to authenticate and identify subscribers on the network. The most important of these are the ICCID, IMSI, Authentication Key (Ki), Local Area Identity (LAI) and Operator-Specific Emergency Number. The SIM also stores other carrier-specific data such as the SMSC (Short Message Service Center) number, Service Provider Name (SPN), Service Dialing Numbers (SDN), Advice-Of-Charge parameters and Value Added Service (VAS) applications. (Refer to GSM 11.11)

SIM cards can come in at least two capacity types: 32kb and 64kb. Both allow a maximum of 250 contacts to be stored on the SIM, but while the 32kb has room for 33 mobile network codes (MNCs) or "network identifiers", the 64kb version has room for 80 MNCs. This is used by network operators to store information on preferred networks, mostly used when the SIM is not in its 'home' country but is roaming. The network operator that issued the SIM card can use this to have a SIM card connect to a preferred network in order to make use of the best price and/or quality network instead of having to pay the network operator and accept its quaility the SIM card simply 'saw' first. This does not mean that a SIM card can only connect to a maximum of 33 or 80 networks, this means that the SIM card issuer can only specify up to that amount of preferred networks, if a SIM is outside these preferred networks it will use the first or best available network.

ICCID

Each SIM is internationally identified by its integrated circuit card identifier (ICCID). ICCIDs are stored in the SIM cards and are also engraved or printed on the SIM card body during a process called personalization. The ICCID is defined by the ITU-T recommendation E.118 as the *Primary Account Number*.[4] Its layout is based on ISO/IEC 7812. According to E.118, the number is up to 19 digits long, including a single check digit calculated using the Luhn algorithm. However, the GSM Phase 1[5] defined the ICCID length as 10 octets with operator-specific structure.

The number is composed of the following subparts:

Issuer identification number (IIN)

Maximum of seven digits:

- Major industry identifier (MII), 2 fixed digits, *89* for telecommunication purposes.
- Country code, 1-3 digits, as defined by ITU-T recommendation E.164.
- Issuer identifier, 1-4 digits.

Individual account identification

- Individual account identification number. Its length is variable, but every number under one IIN will have the same length.

Check digit

- Single digit calculated from the other digits using the Luhn algorithm.

With the GSM Phase 1 specification using 10 octets into which ICCID is stored as packed BCD, the data field has room for 20 digits with hexadecimal digit "F" being used as filler when necessary.

In practice, this means that on GSM SIM cards there are 20-digit (19+1) and 19-digit (18+1) ICCIDs in use, depending upon the issuer. However, a single issuer always uses the same size for its ICCIDs.

To confuse matters more, SIM factories seem to have varying ways of delivering electronic copies of SIM personalization datasets. Some datasets are without the ICCID checksum digit, others are with the digit.

As required by E.118, The ITU regularly publishes a list of all internationally assigned IIN codes in its Operational Bulletins. The most recent list, as of 23 December, 2011, is in Operational Bulletin No. 971 [6]

International mobile subscriber identity (IMSI)

SIM cards are identified on their individual operator networks by a unique IMSI. Mobile operators connect mobile phone calls and communicate with their market SIM cards using their IMSIs. The format is:

- The first 3 digits represent the Mobile Country Code (MCC).
- The next 2 or 3 digits represent the mobile network code (MNC). 3-digit MNC codes are allowed by E.212 but are mainly used in the United States and Canada.
- The next digits represent the mobile station identification number. Normally there will be 10 digits but would be fewer in the case of a 3-digit MNC or if national regulations indicate that the total length of the IMSI should be less than 15 digits.

Authentication key (K_i)

The K_i is a 128-bit value used in authenticating the SIMs on the mobile network. Each SIM holds a unique K_i assigned to it by the operator during the personalization process. The K_i is also stored in a database (termed authentication center or AuC) on the carrier's network.

The SIM card is designed not to allow the K_i to be obtained using the smart-card interface. Instead, the SIM card provides a function, *Run GSM Algorithm*, that allows the phone to pass data to the SIM card to be signed with the K_i. This, by design, makes usage of the SIM card mandatory unless the K_i can be extracted from the SIM card, or the carrier is willing to reveal the K_i. In practice, the GSM cryptographic algorithm for computing SRES_2 (see step 4, below) from the K_i has certain vulnerabilities[7] that can allow the extraction of the K_i from a SIM card and the making of a duplicate SIM card.

Authentication process:

1. When the Mobile Equipment starts up, it obtains the International Mobile Subscriber Identity (IMSI) from the SIM card, and passes this to the mobile operator requesting access and authentication. The Mobile Equipment may have to pass a PIN to the SIM card before the SIM card will reveal this information.
2. The operator network searches its database for the incoming IMSI and its associated K_i.
3. The operator network then generates a Random Number (RAND, which is a nonce) and signs it with the K_i associated with the IMSI (and stored on the SIM card), computing another number known as Signed Response 1 (SRES_1).
4. The operator network then sends the RAND to the Mobile Equipment, which passes it to the SIM card. The SIM card signs it with its K_i, producing SRES_2, which it gives to the Mobile Equipment along with encryption key K_c. The Mobile Equipment passes SRES_2 on to the operator network.

5. The operator network then compares its computed SRES_1 with the computed SRES_2 that the Mobile Equipment returned. If the two numbers match, the SIM is authenticated and the Mobile Equipment is granted access to the operator's network. K_c is used to encrypt all further communications between the Mobile Equipment and the network.

Location area identity

The SIM stores network state information, which is received from the Location Area Identity (LAI). Operator networks are divided into Location Areas, each having a unique LAI number. When the device changes locations, it stores the new LAI to the SIM and sends it back to the operator network with its new location. If the device is power cycled, it will take data off the SIM, and search for the prior LAI. This saves time by avoiding having to search the whole list of frequencies that the telephone normally would.

SMS messages and contacts

Most SIM cards will orthogonally store a number of SMS messages and phone book contacts. The contacts are stored in simple 'Name and number' pairs: entries containing multiple phone numbers and additional phone numbers will usually not be stored on the SIM card. When a user tries to copy such entries to a SIM the handset's software will break them up into multiple entries, discarding any information that isn't a phone number. The number of contacts and messages stored depends on the SIM; early models would store as few as 5 messages and 20 contacts while modern SIM cards can usually store over 250 contacts.

Formats

SIM cards are available in four standard sizes; full-size, mini-SIM, micro-SIM and embedded SIM. The first to appear was the full-size and is the size of a credit card (85.60 mm × 53.98 mm × 0.76 mm). A newer, more popular, version has the same thickness but has a length of 25 mm and a width of 15 mm, and has one of its corners truncated (chamfered) to prevent misinsertion. The newest incarnation, known as the micro-SIM or 3FF, has dimensions of 15 mm × 12 mm.

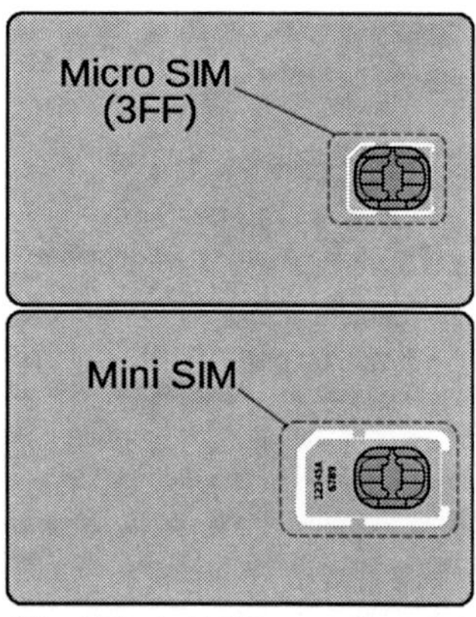

Micro-SIM and mini-SIM, as normally supplied in full-sized carrier cards.

The mini-SIM card has the same contact arrangement as the full-size SIM card and is normally supplied within a full-size card carrier, attached by a number of linking pieces. This arrangement (defined in ISO/IEC 7810 as ID-1/000) allows for such a card to be used in a device requiring a full-size card, or to be used in a device requiring a mini-SIM card after cleanly breaking the scorings manufactured in the outline of a mini-SIM card.

For use in even smaller devices, the *3FF card* or *micro-SIM* cards have the same thickness and contact arrangements, but the length and width are further reduced to 15 mm × 12 mm. The specifications for the 3FF card or micro-SIM also include additional functionality beyond changing the physical card size.

SIM cards forM2M applications are available in a surface mount SON-8 package which may be soldered directly onto a circuit board.

Micro-SIM with mini-SIM and full SIM brackets from Telia in Sweden

The memory film from a micro SIM card without the plastic backing plate, next to a US dime, which is approx. 18 mm in diameter.

Embedded SIM from M2M supplier Eseye with an adapter board for evaluation in a Mini-SIM socket

SIM card sizes

SIM card	Standard reference	Length (mm)	Width (mm)	Thickness (mm)
Full-size	ISO/IEC 7810:2003, ID-1	85.60	53.98	0.76
Mini-SIM	ISO/IEC 7810:2003, ID-000	25.00	15.00	0.76
Micro-SIM	ETSI TS 102 221 V9.0.0, Mini-UICC	15.00	12.00	0.76
Embedded-SIM	JEDEC Design Guide 4.8 , SON-8	6.00	5.00	<1.0

The micro-SIM was developed by the European Telecommunications Standards Institute (ETSI) along with SCP, 3GPP (UTRAN/GERAN), 3GPP2 (CDMA2000), ARIB, GSM Association (GSMA SCaG and GSMNA), GlobalPlatform, Liberty Alliance, and the Open Mobile Alliance (OMA) for the purpose of fitting into devices otherwise too small for a mini-SIM card.[2] [8]

The form factor was mentioned in the Dec 1998 3GPP SMG9 UMTS Working Party, which is the standards-setting body for GSM SIM cards,[9] and the form factor was agreed upon in late 2003.[10]

The micro-SIM was created for backward compatibility. The major issue with backward compatibility was the contact area of the chip. Retaining the same contact area allows the micro-SIM to be compatible with the prior, larger SIM readers through the use of plastic cutout surrounds. The SIM was also designed to run at the same speed (5 MHz) as the prior version. The same size and positions of pins resulted in numerous "How-to" tutorials and YouTube video with detailed instructions how to cut a mini-SIM card to micro-SIM size with a sharp knife or scissors. These tutorials became very popular among first owners of iPad 3G after its release on April 30, 2010 and iPhone 4 on June 24, 2010.[11]

The chairman of EP SCP, Dr. Klaus Vedder, said[10]

> "With this decision, we can see that ETSI has responded to a market need from ETSI customers, but additionally there is a strong desire not to invalidate, overnight, the existing interface, nor reduce the performance of the cards. EP SCP expect to finalise the technical realisation for the third form factor at the next SCP plenary meeting, scheduled for February 2004."

The surface mount format provides the same electrical interface as the Full size, 2FF and 3FF SIM cards, but is soldered to the circuit board as part of the manufacturing process. In M2M applications where there is no requirement to change the SIM card, this avoids the requirement for a connector, improving reliability and security.

Developments of SIM

A *virtual SIM* is a mobile phone number provided by a mobile network operator that does not require a SIM card to connect phone calls to a user's mobile phone.

USIM (Universal Subscriber Identity Module) is an application for UMTS mobile telephony running on a UICC smart card which is inserted in a 3G mobile phone. There is a common misconception to call the UICC itself a USIM, but the USIM is merely a logical entity on the physical card. It stores user subscriber information, authentication information and provides storage space for text messages and phone book contacts. The phone book on a UICC has been greatly enhanced. For authentication purposes, the USIM stores a long-term pre-shared secret key, which is shared with the Authentication Center (AuC) in the network. The USIM also verifies a sequence number that must be within a range using a window mechanism to avoid replay attacks, and is in charge of generating the session keys to be used in the confidentiality and integrity algorithms of the KASUMI block cipher in UMTS. The equivalent of USIM on CDMA networks is CSIM.

Usage in mobile phone standards

The use of SIM cards is mandatory in GSM devices.

The satellite phone networks Iridium, Thuraya and Inmarsat's BGAN also use SIM cards. Sometimes these SIM cards work in regular GSM phones and also allow GSM customers to roam in satellite networks by using their own SIM card in a satellite phone.

SIM card for Thuraya satellite phone

Japan's 2G PDC system (which will be completely shut down by 2012; SoftBank Mobile has already shut down PDC from March 31, 2010) also specifies a SIM, but this has never been implemented commercially. The specification of the interface between the Mobile Equipment and the SIM is given in the RCR STD-27 annex 4. The Subscriber Identity Module Expert Group was a committee of specialists assembled by the European Telecommunications Standards Institute (ETSI) to draw up the specifications (GSM 11.11) for interfacing between smart cards and mobile telephones. In 1994, the name SIMEG was changed to SMG9.

Grameenphone's SIM card

Japan's current and next generation cellular systems are based on W-CDMA (UMTS) and CDMA2000 and all use SIM cards.

CDMA-based devices originally did not use a removable card, and the service for these phones bound to a unique identifier contained in the handset itself. This is most prevalent in operators in the Americas. The first publication of the TIA-820 standard (also known as 3GPP2 C.S0023) in 2000 defined the Removable User Identity Module (R-UIM). Card-based CDMA devices are most prevalent in Asia.

The equivalent of a SIM in UMTS is called the Universal Integrated Circuit Card (UICC), which runs a USIM application. The UICC is still colloquially called a *SIM card*.

KDDI's au IC-Card

SIM and carriers

The SIM card introduced a new and significant business opportunity of mobile telecommunications operator/carrier business of the mobile virtual network operator (MVNO) which does not own or operate a cellular telecoms network, but which leases capacity from one of the network operators, and only provides a SIM card to its customers. MVNOs first appeared in Denmark, Hong Kong, Finland and the UK and today exist in over 50 countries, including most of Europe, United States, Canada, Australia and parts of Asia, and account for approximately 10% of all mobile phone subscribers around the world.

On some networks, the mobile phone is locked to its carrier SIM card, meaning that only the specific carrier's SIM cards will work. This is more common in markets where mobile phones are heavily subsidised by the carriers, and the business model depends on the customer staying with the service provider for a minimum term (typically 12 or 24 months). Common examples are the GSM networks in the United States, Canada, Australia, the UK and Poland. Many businesses offer the ability to remove the SIM lock from a phone, effectively making it possible to then use the phone on any network by inserting a different SIM card. Mostly, GSM and 3G mobile handsets can easily be unlocked and used on any suitable network with any SIM card.

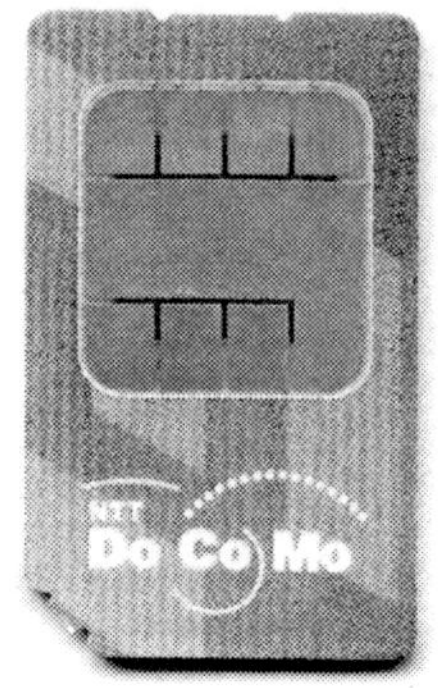

NTT DoCoMo's FOMA Card

In countries where the phones are not subsidised, *e.g.*, Italy, India and Belgium, all phones are unlocked. Where the phone is not locked to its SIM card, the users can easily switch networks by simply replacing the SIM card of one network with that of another while using only one phone. This is typical, for example, among users who may want to optimise their carrier's traffic by different tariffs to different friends on different networks, or when traveling internationally.

SIMware

A term used to describe software held on or running on a SIM card.[12]

See also

- International Mobile Equipment Identity (IMEI)
- GSM 03.48
- SIM cloning
- SIM connector
- IP Multimedia Services Identity Module (ISIM)
- W-SIM (Willcom-SIM)
- Mobile equipment identifier (MEID)
- VMAC
- Mobile signature
- Single Wire Protocol (SWP)
- SIM Application Toolkit (STK)
- Mobile broadband
- Tethering
- Smart card

References

[1] History of Giesecke & Devrient (http://www.gi-de.com/portal/page?_pageid=44,62528&_dad=portal&_schema=PORTAL)
[2] Gaby Lenhart, Project leader: ETSI Technical Committee Smart Card Platform (TB SCP) (2006-04-01). "The Smart Card Platform" (http://docbox.etsi.org//Workshop/2006/Salud Mexico/Gaby Lenhart - CENETEC_2006_04.ppt). ETSI Technical Committee Smart Card Platform (TB SCP). . Retrieved 30 January 2010. "SCP is co-operating on both technical and service aspects with a number of other committees both within and outside the telecommunications sector."
[3] "The Java Card 3 Platform" (http://java.sun.com/javacard/3.0/javacard3_whitepaper.pdf). *August 2008*. Sun Microsystems. . Retrieved 13 August 2011.
[4] ITU-T, ITU-T Recommendation E.118, The international telecommunication charge card, Revision history (http://www.itu.int/rec/T-REC-E.118), Revision "05/2006" (http://www.itu.int/rec/dologin_pub.asp?lang=e&id=T-REC-E.118-200605-I!!PDF-E&type=items)
[5] ETSI, ETSI Recommendation GSM 11.11, Specifications of the SIM-ME Interface, Version 3.16.0 (http://www.3gpp.org/ftp/Specs/archive/11_series/11.11/1111-3G0.ZIP)
[6] http://www.itu.int/pub/T-SP-OB.971-2011
[7] "Hackers crack open mobile network" (http://www.bbc.co.uk/news/technology-13013577). *31 December 2010* (bbc.co.uk). 20 April 2011. . Retrieved 13 August 2011.
[8] Segan, Sascha (2010-01-27). "Inside the iPad Lurks the 'Micro SIM'" (http://www.pcmag.com/article2/0,2817,2358489,00.asp). PC Magazine. . Retrieved 30 January 2010.
[9] "DRAFT Report of the SMG9 UMTS Working Party, meeting #7 hosted by Nokia in Copenhagen, 15.-16- December 1998" (http://www.3gpp.org/ftp/TSG_T/WG3_USIM/TSGT3_01/docs/t3-99003.pdf). *3GPP*. 25 January 1999. . Retrieved 27 January 2010. "One manufacturer stated that it may be difficult to meeting ISO mechanical standards for a combined ID-1/micro-SIM card."
[10] Antipolis, Sophia (2003-12-08). "New form factor for smart cards introduced" (http://www.smartcardstrends.com/det_atc.php?idu=287). SmartCard Trends. . Retrieved 30 January 2010. "The work item for the so-called Third Form Factor, "3FF", was agreed, after intensive discussions, at the SCP meeting held last week in London."
[11] "How to make MicroSIM" (http://www.youtube.com/watch?v=6B0G_qjebJY). .
[12] "Case details for Trade Mark 2516989" (http://www.ipo.gov.uk/domestic?domesticnum=2516989), Intellectual Property Office (UK)

Further reading

- Sim Card Cloning (http://www.5ne.org/sim-cloning)

External links

- ETSI Smart Card standards (102 221) (http://pda.etsi.org/pda/AQuery.asp?qSEARCH_STRING=102+221&qSEARCH_Type=EXACT&submit1=Search&qNB_TO_DISPLAY=10&qOFFSET=0&qSORT=DEFAULT)
- GSM 11.11 (http://www.3gpp.org/ftp/specs/html-info/1111.htm) - Specification of the Subscriber Identity Module — Mobile Equipment (SIM — ME) interface.
- GSM 11.14 (http://www.3gpp.org/ftp/specs/html-info/1114.htm) - Specification of the SIM Application Toolkit for the Subscriber Identity Module — Mobile Equipment (SIM — ME) interface
- GSM 03.48 (http://www.3gpp.org/ftp/specs/html-info/0348.htm) - Specification of the security mechanisms for SIM application toolkit.
- GSM 03.48 Java API (http://gsm0348.googlecode.com) - API and realization of GSM 03.48 in Java.
- ITU-T E.118 (http://www.itu.int/rec/T-REC-E.118-200605-I/en) - The International Telecommunication Charge Card. 2006 ITU-T.

HTC_Flyer

HTC Flyer (Front Cover View)

Manufacturer	HTC Corporation
Type	Tablet, media player
Generation	Initial Release
Release date	May 2011
Introductory price	Best Buy: US$499.99,[1] Amazon.com: US$794.99[2]
Operating system	Android 3.2.1 with HTC Sense UI [3] [4]
Power	Battery/AC
CPU	1.5 GHz Qualcomm Snapdragon
Storage capacity	Flash memory eMMC: 16 and 32 GB
Memory	1 GB DDR2 RAM
Display	7-inch 1024×600 px
Input	Multi-touch capacitive touchscreen display HTC Scribe Capacitive Stylus
Camera	5.0 megapixel rear-facing 1.3 megapixel, front-facing
Connectivity	HSDPA 900 / 1700 / 2100 Wi-Fi 802.11b/g/n Bluetooth 3 (with A2DP)[3]
Dimensions	195.4 mm (7.69 in) *(h)* 122 mm (4.8 in) *(w)* 13.2 mm (0.52 in) *(d)*
Weight	415 g (14.6 oz)
Website	HTC Flyer [5] at HTC.com

The **HTC Flyer** (also known as the **HTC EVO View 4G**) is a tablet computer by HTC Corporation. It was announced at the Mobile World Congress (MWC) 2011 and released in May 2011.[6] Unlike other tablets announced at MWC, the Flyer has a single-core 1.5 GHz CPU[7] and ran 2.3.3 (Gingerbread). .[3] [4] In February 2011 it was reported that HTC had claimed via Facebook that "Flyer will be getting a Honeycomb upgrade in Q2", however an HTC representative subsequently stated "I can confirm that we are working to bring a Honeycomb update to Flyer in short order – however, I don't have any specific information on what the timing may be."[8] The version running Android 3.2.1 (Honeycomb) was released later in 2011.

It has a 7 inch TFT display and includes some special features, such as pen input as well as touch input.[9] [10] [11] A Wi-Fi variant is expected to be launched at a lower price soon after the 3G variant is released.[12]

Key features

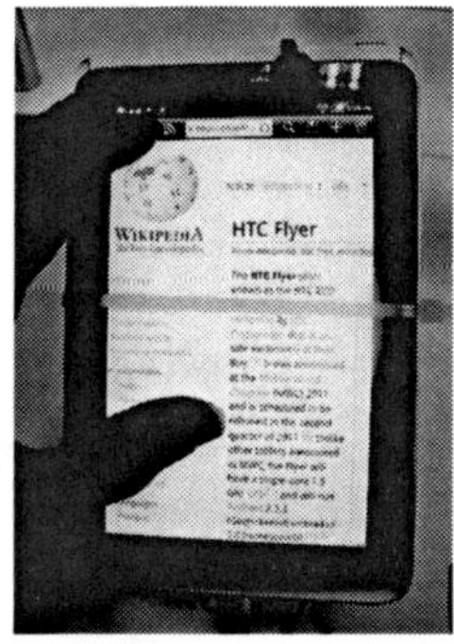
In use

HTC Flyer- Back

- An optional HTC Scribe digital pen using HTC Scribe technology running on top of N-trig DuoSense hardware, by which a user can capture and annotate any on-screen content with notes and drawings using a battery powered, active, non-capacitive digital stylus.
- 7" 1024×600 display with multi-touch capability
- Integration and compatibility with Wi-Fi and Bluetooth 3.0
- Built-in dual microphones for noise reduction
- Android operating system
- Adobe Flash 10.3 and HTML5 support

Hardware

- 5 MP Color CMOS camera with auto focus
- 1.3MP front camera for video chatting
- 16GB/32GB eMMC memory plus microSD card slot
- 4,000 mAh battery

"The HTC Flyer features an immersive 3D user experience that brings to life, all your favorite content - weather, email and even ebooks."[13] Best Buy was the exclusive retailer for the spring 2011 launch.[14] Currently, Sprint will be the only carrier to sell a 4G model with black and red instead of white and silver.

References

[1] "Best Buy - HTC Flyer Tablet (White)" (http://www.bestbuy.com/site/HTC+-+Flyer+Tablet+-+White/2390524.p?id=1218332598563&skuId=2390524). .

[2] "Amazon.com: HTC Flyer 3G/WiFi" (http://www.amazon.com/dp/B004VF612G). .

[3] Bray, Jonathan (15 February 2011). "HTC Flyer review: first look" (http://www.pcpro.co.uk/blogs/2011/02/15/htc-flyer-review-first-look/). PC Pro UK. . Retrieved 17 February 2011.

[4] "Version confusion on Google's Android resolved" (http://www.h-online.com/open/news/item/Version-confusion-on-Google-s-Android-resolved-1195322.html). 23 February 2011. . Retrieved 6 March 2011.

[5] http://www.htc.com/www/product/flyer/overview.html

[6] "HTC Flyer Tablet launched at Mobile World Congress 2011" (http://www.newsden.net/htc-flyer-tablet-specs-review-hands-on-video-6660/). .

[7] Palmer, Jacob (17 February 2011). "LG Optimus Pad vs HTC Flyer vs Samsung Galaxy Tab 2" (http://www.gizmocrunch.com/computing/5391-htc-flyer-lg-optimus-pad-samsung-galaxy-tab-2). Gizmocrunch. . Retrieved 17 February 2011.

[8] "Honeycomb to come to HTC Flyer" (http://www.androidcentral.com/honeycomb-come-htc-flyer-q2). Android Central. 15 February 2011. .

[9] Kremp, Matthias (17 February 2011). "Schlaumeier-Handys und Facebook überall" (http://www.spiegel.de/netzwelt/gadgets/0,1518,745673,00.html) (in German). Der Spiegel. . Retrieved 17 February 2011.

[10] Briggeman, Mark (15 February 2011). "Hands-on with the new HTC Flyer" (http://www.mobilityminded.com/12194/hands-on-with-the-new-htc-flyer). Mobility Minded. . Retrieved 17 February 2011.

[11] "MWC2011: HTC Flyer Android 2.4 Tablet with 1 GB RAM and 1.5 GHz processor" (http://mobilesmug.com/news/49-tablets/785-mwc2011-htc-flyer-android-24-tablet-with-1-gb-ram-and-15-ghz-processor-release-date-available). .
[12] "HTC Flyer set for cheaper Wi-Fi option News TechRadar UK" (http://www.techradar.com/news/mobile-computing/htc-flyer-set-for-cheaper-wi-fi-option-928670). .
[13] HTC Flyer coming to Singapore in May (http://www.vr-zone.com/articles/htc-flyer-coming-to-singapore-in-may/11968.html)
[14] "Best Buy Gets Exclusive Agreement to Sell HTC Flyer" (http://www.htctablet.net/2011/best-buy-gets-exclusive-agreement-to-sell-htc-flyer/). *HTCFlyer.net*. March 23, 2011.

External links

- Official website (http://www.htc.com/www/product/flyer/overview.html)
- Official website: product specs (http://www.htc.com/www/product/flyer/specification.html)
- HTC Flyer Tablet community (http://htcflyertablet.com)

Crayon Physics Deluxe

Crayon Physics	
Developer(s)	Petri Purho
Designer(s)	Petri Purho
Composer(s)	Stian Stark
Engine	Box2D
Platform(s)	Windows, iOS, Mac OS X, Linux
Release date(s)	January 7, 2009
Genre(s)	Video puzzle game
Mode(s)	Single-player
Rating(s)	• Apple: 4+

Crayon Physics Deluxe is a puzzle game designed by Petri Purho and released on January 7, 2009. An early version, titled ***Crayon Physics***, won the grand prize at the Independent Games Festival in 2008. It features a heavy emphasis on two-dimensional physics simulations, including gravity, mass, kinetic energy and transfer of momentum. The game includes a level editor and enables its players to share and download custom content via an online service.

Gameplay

The objective of each level in *Crayon Physics Deluxe* is to guide a ball from a predetermined start point so that it touches all of the stars placed on the level. The ball and nearly all objects on the screen are affected by gravity. The player cannot control the ball directly, but rather must influence the ball's movement by drawing physical objects on the screen. Depending on how the object is drawn, it becomes a rigid surface, a pivot point, a wheel or a rope, and the object can then interact with the ball by hitting it, providing a surface to roll on, dragging, carrying or launching the ball, etc. The player can also nudge the ball left or right by clicking on it, and in some levels, rockets appear and can be used as part of the solution.

The game challenges players to come up with creative solutions to each puzzle, and provides additional rewards for elegant solutions that don't rely on "brute force" methods. It comes with more than seventy levels, and also features a level editor and an online Playground, where players can upload and download custom levels.

Development

Crayon Physics

Crayon Physics, the original prototype of this game, is Purho's tenth "rapid-prototype project" inspired by the rules of the Experimental Gameplay Project, and was developed in five days[1] using resources freely available under the Creative Commons license. The game was inspired by descriptions Purho had heard of the children's book *Harold and the Purple Crayon*. On June 10, 2007, Purho announced that he would be developing a level editor to permit user-created levels, although by June 15 fans of the game had already worked out the level format and had released new levels for the game. The level editor was released on June 30.

Crayon Physics Deluxe

On October 12, 2007, Purho announced *Crayon Physics Deluxe*, which would feature an intuitive level editor, more levels, and a modification to the game engine to preserve the player's drawings instead of turning them into rectangles.[1] The follow-up took a year and eight months to develop.[2] It won the Seumas McNally Grand Prize at the Independent Games Festival in February 2008.[3] Chris Baker of *Slate Magazine* also wrote that *Crayon Physics Deluxe* was more talked about than *Gears of War 2* at the 2008 Game Developers Conference.[1]

Platforms

Published by Hudson Soft, *Crayon Physics Deluxe* was released for the iOS on January 1, 2009 and in Spring 2010 for the iPad via Apple's App Store.[4] A version for the PC was released six days later.[5] An unofficial clone was made for the DS, but only in free play mode and under the title of *Pocket Physics*[6] . A port for Windows Mobile was also made, but later pulled. It can still be downloaded unofficially. [7]

See also

- Microsoft Physics Illustrator

References

[1] Chris Baker (2008-03-19). "Crayon Physics Deluxe, an ingenious video game that looks like it was designed by a third-grader" (http://slate.com/id/2186848/). Slate (magazine). pp. 2. . Retrieved 2008-03-20.

[2] "Computer Game A Mash-Up Of Crayons, Physics" (http://npr.org/templates/story/story.php?storyId=99080116). All Things Considered. 9 January 2009. .

[3] "2008 Independent Games Festival Winners" (http://igf.com/02finalists.html). *Independent Games Festival*. Think Services. . Retrieved 2008-10-13.

[4] "Day 1: Crayon Physics Deluxe for the iPhone" (http://kloonigames.com/blog/crayonphysics/day-1-crayon-physics-deluxe-for-the-iphone). *Kloonigames*. Petri Purho. 2009-01-01. . Retrieved 2009-01-04.

[5] "Day 7: It's Here!" (http://kloonigames.com/blog/crayonphysics/day-7-its-here/). *Kloonigames*. Petri Purho. 2009-01-07. . Retrieved 2009-01-08.

[6] "Pocket Physics" (http://tobw.net/index.php?cat_id=3&project=Pocket+Physics). Tobias Weyand. .

[7] "Crayon Physics Game Pulled from Samsung Omnia II" (http://pocketnow.com/software-1/crayon-physics-game-pulled-from-samsung-omnia-ii). *Pocketnow.com*. CJ Lippstreu. 2010-01-01. .

External links

- Crayon Physics Deluxe (http://crayonphysics.com); offers a free demo and the full game for sale
- Gameplay of Crayon Physics Deluxe using a tablet PC's touchscreen (http://youtube.com/watch?v=QsTqspnvAaI) on YouTube
- Crayon Physics Deluxe for the Apple iPhone and iPod Touch (http://crayonphysicsdeluxe.com)
- Crayon Physics (http://kloonigames.com/blog/games/crayon) free prototype

Wacom

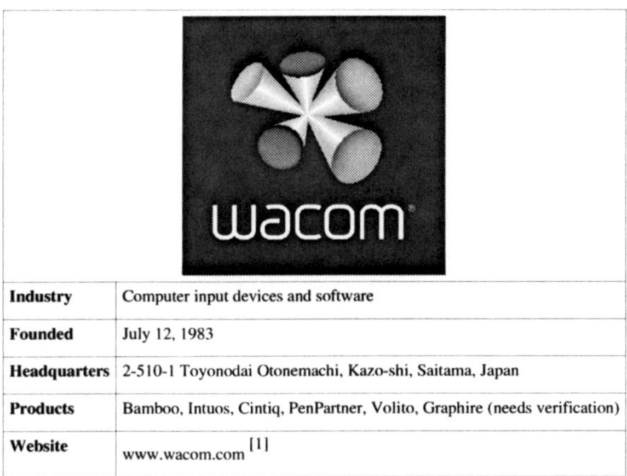

Industry	Computer input devices and software
Founded	July 12, 1983
Headquarters	2-510-1 Toyonodai Otonemachi, Kazo-shi, Saitama, Japan
Products	Bamboo, Intuos, Cintiq, PenPartner, Volito, Graphire (needs verification)
Website	www.wacom.com [1]

Wacom Co., Ltd. (*Kabushiki-gaisha Wakomu*) (English pronunciation: /ˈwɑːkəm or ˈweɪˈkɔm/[2] (TYO: 6727 [3]) is a worldwide company that produces graphics tablets and related products, headquartered in Otone, Saitama, Japan. The American headquarters are located in Vancouver, Washington, and those for EMEA (Europe, Middle East and Africa) in Krefeld, Germany. **Wacom** is a Japanese portmanteau: *Wa* for "harmony" or "circle", and *Komu* for "computer". Wacom tablets are notable for their use of a patented cordless, battery-free, and pressure-sensitive stylus or digital pen. In addition to manufacturing and selling tablets as separate products, Wacom also provides graphical input technology for some tablet computers, which it calls "Penabled Technology".

The Wacom tablet functionality was in 1992 used in the screen of the Compaq Concerto computer, making it into an early tablet computer.

Market share

Year	Japan	Rest of the World	Notes
2005	95.8%	70%	For the number for Japan, "Survey by BCN Research Institute in 2005" is cited.[4]
2008	95.4%	86%	For the number for Japan, Survey by BCN Research Inc. in 2008 was cited.[5]
2009	93.8%	85%	For the number for Japan, "Survey by BCN Inc. in 2009" is cited[6]
2010	85.7%	85%	For the number for Japan, "Survey by BCN Inc. in 2010" is cited[7]

Product lines

Wacom produces several lines of tablets, three of which are marketed worldwide. Most tablets are sold with a bundle of software such as Corel Painter Essentials and Photoshop Elements, which take advantage of the features of the tablet. Each is sold with a digital pen that is compatible with that model; digital pens generally do not work with tablets of a different product line or generation. Some of these pens include features such as additional buttons on the shaft or an "eraser" at the other end. Some tablet models include a puck (mouse) based on the same technology. Software drivers for recent versions of Mac OS X and Microsoft Windows are included with most models. All current models of external tablet connect to computers via USB or Bluetooth.

Bamboo

The Bamboo line is aimed at home users. Current models† feature 1024 levels of pressure sensitivity and a resolution of 2540 lines per inch[8] [9] [10] [11] (1000 lines/cm). Most of the models have a 5.8 × 3.6 in (14.7 × 9.2 cm) active surface area, while the larger "Fun" and "Create" models have a usable surface area of 8.5 × 5.4 in (22 × 14 cm). The Bamboo One (no longer produced; sold only in Europe) was an approximately A6-sized tablet which used the same eraser-less pen as the other models but, unlike other models, featured no control buttons.

Bamboo tablets feature a battery-free pen (powered by the same EMR technology as the Intuos line), which can be used alongside finger swipes (in some models), with ± 0.02 in (± 0.5 mm) accuracy. The "Pen&Touch" model includes an option to switch orientation for left- or right-handed users.

In the Americas, there are three models currently† available: Bamboo Create, Bamboo Capture and Bamboo Connect.[12] In addition to stylus-based input, "Create" and "Capture" models feature multi-touch functionality, with support for one- and two-finger gestures for such operations as scrolling and zooming. The "Create" model includes an eraser-equipped stylus, and additional bundled graphics software (Adobe Photoshop Elements 7, Corel Painter Essentials 4 and Nik Color Efex Pro 3).

In Europe, there are four tablet models currently† available: Bamboo Pen, Bamboo Pen&Touch, Bamboo Fun Small Pen&Touch and Bamboo Fun Medium Pen&Touch.[11]

Bamboo tablets

Tablet	Region sold	In production†	multi-touch	Active surface area		Aspect ratio	Notes
				Centimeters	Inches		
Create [13]	Americas	Yes	Yes	22 × 14 cm	8.5 × 5.4 in	16:10	
Capture [14]	Americas	Yes	Yes	14.7 × 9.2 cm	5.8 × 3.6 in		
Connect [15]	Americas	Yes	No				
Pen&Touch	Europe	Yes	Yes[16]	14.7 × 9.2 cm for pen input[16] 12.5 × 8.5 cm for touch input[16]	5.8 × 3.6 in for pen input 4.9 × 3.3 in for touch input		
Fun Medium Pen&Touch	Europe	Yes	Yes				
Fun Small Pen&Touch	Europe	Yes	Yes	14.7 × 9.2 cm for pen input 12.5 × 8.5 cm for touch input	5.8 × 3.6 in for pen input 4.9 × 3.4 in for touch input		
One	Europe	No	No	≈ A6		4:3	Does not feature control buttons

† As of October 6, 2011

Intuos

Wacom Intuos4 Medium Pen Tablet with pen.

Intuos is marketed to professional graphic artists, and features the highest specifications of any Wacom device. It has a similar feel to drawing on paper. The latest version, Intuos 4, is available in multiple sizes and proportions, and it includes 60 degrees of tilt sensitivity (50 degrees in small model) and 2048 levels of pressure due to EMR (Electro Magnetic Resonance) technology in the Wacom pen. The Intuos line offers an 5080 lines per inch resolution and comes in the following sizes (active area):

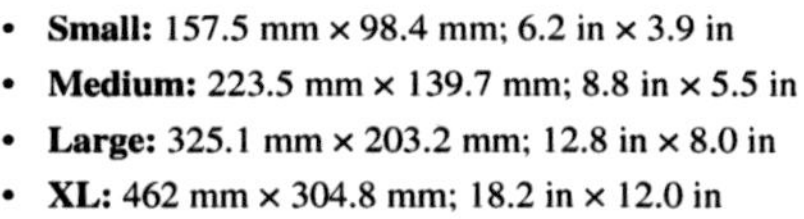

- **Small:** 157.5 mm × 98.4 mm; 6.2 in × 3.9 in
- **Medium:** 223.5 mm × 139.7 mm; 8.8 in × 5.5 in
- **Large:** 325.1 mm × 203.2 mm; 12.8 in × 8.0 in
- **XL:** 462 mm × 304.8 mm; 18.2 in × 12.0 in
- **Wireless:** 203.2 mm × 127 mm; 8.0 in × 5.0 in

All sizes but Small feature 8 ExpressKeys and an OLED display next to them to show the function of each button. The Small sacrifices the OLED display and two ExpressKeys.

Cintiq

The Cintiq is a tablet/screen hybrid, a graphics tablet that incorporates an LCD into the digitizing tablet itself, allowing the user to draw directly "on" the display surface. The tablets are available in several sizes.

A 21-inch 1600×1200 resolution tablet, the 21UX, has been available for several years at various price points. As of November 2007, both a 12-inch and a 20-inch widescreen model were released, the 12WX and the 20WSX, respectively. All three models utilize Intuos3 pen technology with 1024 levels of pressure sensitivity.

In 2008, the 21UX saw a major design revision and a price drop. It has a higher pen resolution and 2048 pressure levels due to Wacom's integration of the Intuos4's technology. The 21UX's sensitivity is much greater than most tablet computers and portable computers offering similar functionality on built-in screens. In addition, the 21UX includes an integrated stand that allows the user to tilt and rotate the unit as he or she prefers.

On September 13, 2011, Wacom announced its newest and largest Cintiq, the 24HD (DTK-2400). It contains a 24-inch 1920×1200 resolution LCD with 92% coverage of the Adobe RGB color gamut (versus 72% in the 21UX). The pen technology, like the 21UX, is identical to the Intuos4 in terms of resolution per inch and sensitivity. This model includes an integrated adjustable-tilt stand that allows the 24HD to hang off the edge of the table and closer to the user.

Inkling

Inkling, announced on August 30, 2011, is a new device that enables artists to draw sketches on paper that can then be converted into digital images. Inkling consists of a receiver, which artists insert any kind of paper into, and a special pen which uses real ink. As artists draw on paper, they are able to add new layers by tapping a button on the receiver. When the artist have finished their sketches, they can connect the receiver into a USB port, where the sketches are imported into Sketch Manager and can be exported to Photoshop, Illustrator or Autodesk Sketchbook Pro, as well as various image formats. The device was originally due to go on sale mid-September, 2011, priced at around US$199/€169. However, its release was later pushed back to September 30, 2011; then again to mid-October, 2011; then again to mid-November, 2011; and then again to an unknown date.[17] [18]

Other lines

Wacom has additional products which it markets in various parts of the world. The "Graphire Wireless" is a 6 × 8 in (15 × 20 cm) version of Wacom's discontinued Graphire line (largely replaced by the Bamboo) which communicates with the computer via Bluetooth rather than a USB cable. It remains available in the Americas. The Graphire digital pen is interchangeable with the original Bamboo model's digital pen but not later Bamboos. In Europe, Wacom offers the "Colorelli", a tablet and software package marketed as a creative outlet for children; "JustWrite Office" a basic tablet for capturing written input in office applications; the "PL Series", similar in function to the Cintiq but with more modest specifications suited for office use; and the "Signature Tablet", a monochrome display/tablet for capturing signatures.

Discontinued

Previous products from Wacom included the ArtZ, ArtZ II, ArtPad, ArtPad II, Digitizer, Digitizer II, Favo, UltraPad, Graphire through Graphire4, Intuos through Intuos3, 15-, 17- and 18-inch Cintiqs, Volito, and PenPartner. Early models used RS-232 and Apple serial connectors, with a conversion to USB in later models.

Criticism

Several Wacom models, including the Intuos4 and Bamboo, have been criticized for the drawing surface's roughness, which rapidly wears down nibs and can result in uneven wear patterns, leaving slick and non-slick areas. Fortunately, being made of nothing more than plastic, nibs can be replaced by a short length of nylon 'wire' (approx 0.065 inches or 1.7mm diameter) like that found in grass trimmer or 'weed-eater' refills, suitably straightened out by hand and smoothed (rounded off) at one end with abrasive paper.[19] [20] [21] [22] [23] [24] [25]

The Intuos 4 surface sheet was revised in October 2010 to reduce nib wear. Wacom Europe sells a bundle that includes the revised surface sheet and replacement nibs at a reduced price for installation in existing Intuos 4 tablets.[26]

Drivers

Wacom supplies drivers for contemporary versions of Microsoft Windows and Mac OS. The driver package includes a control panel which allows extensive customization of how the tablet and pen work with the host OS.

The Linux Wacom Project [27] produces drivers for Linux/X11, and is maintained by a Wacom employee.

ThinkyHead Software publishes the free TabletMagic [28] driver package. TabletMagic is a driver for discontinued (serial port) Wacom tablets for use on modern Apple Macintosh computers under the Mac OS X operating system. A USB-to-serial port adapter is required. (OS X open source drivers for many such adapters are available from Source Forge [29].) Not all original functions of the tablet are supported by TabletMagic, but most basic functions are retained.

Technology

Wacom tablets use a patented electromagnetic resonance technology.[30] Since the tablet provides power to the pen through resonant coupling, no battery or cord is required for the pointing device. As a result, there are no batteries inside the pen (or the accompanying puck), which makes them slimmer.

Under the tablet's surface (or LCD in the case of the Cintiq) is a printed circuit board with a grid of multiple send/receive coils and a magnetic reflector attached behind the grid array. In send mode, the tablet generates a close-coupled electromagnetic field (also known as a B-field) at a frequency of 531 kHz. This close-coupled field stimulates oscillation in the pen's coil/capacitor (LC) circuit when brought into range of the B-field. Any excess resonant electromagnetic energy is reflected back to the tablet. In receive mode, the energy of the resonant circuit's oscillations in the pen is detected by the tablet's grid. This information is analyzed by the computer to determine the pen's position, by interpolation and Fourier analysis of the signal intensity. In addition, the pen communicates other vital information, such as pen tip pressure, side-switch status, tip vs. eraser orientation, and the ID number of the tool (to differentiate between different pens. mice, etc.). For example, applying more or less pressure to the tip of the pen changes the value of the pen's timing circuit capacitor. This signal change can be communicated in an analog or digital method. An analog implementation would modulate the phase angle of the resonant frequency, and a digital method is communicated to a modulator which distributes the information digitally to the tablet. The tablet forwards this and other relevant tool information in packets, up to 200 times per second, to the computer.

References

[1] http://www.wacom.com/

[2] "Wacom Asia Pacific | Vista" (http://www.wacom-asia.com/vista/index.html). . Retrieved 2010-09-23.

[3] http://www.bloomberg.com/apps/quote?ticker=6727:JP

[4] "FAQs" (http://web.archive.org/web/20071005131703/http://www.wacom.com/ir/faq.html). Wacom. Archived from the original (http://www.wacom.com/ir/faq.html) on 2007-10-05. . Retrieved 2008-02-16.

[5] "FAQs" (http://www.wacom.com/ir/faq.html). Wacom. . Retrieved 2009-08-14.

[6] "FAQs" (http://www.wacom.co.jp/corporate/en/ir/faq.html). Wacom. . Retrieved 2010-07-10.

[7] "FAQs" (http://www.wacom.co.jp/corporate/en/ir/faq.html#24). Wacom. . Retrieved 2011-08-06.

[8] "Bamboo Create" (http://www.wacom.com/en/Products/Bamboo/~/link.aspx?_id=2E1CA7204E054F39B37D9F404B5C593D&_z=z). Wacom. .

[9] "Bamboo Connect" (http://www.wacom.com/en/Products/Bamboo/~/link.aspx?_id=877DF9CCA96144F0BA0332CED36EF468&_z=z). Wacom. .

[10] "Bamboo Capture" (http://www.wacom.com/en/Products/Bamboo/~/link.aspx?_id=1D60C6DDB67A4A01B79A81AD39C3EDB5&_z=z). Wacom. .

[11] "Compare Bamboo tablets" (http://www.wacom.eu/index2.asp?pid=9240&spid=8&lang=en). Wacom Europe GmbH. .

[12] "Bamboo Tablets" (http://www.wacom.com/en/Products/Bamboo/BambooTablets.aspx). Wacom. .

[13] http://www.wacom.com/en/Products/Bamboo/~/link.aspx?_id=2E1CA7204E054F39B37D9F404B5C593D&_z=z

[14] http://www.wacom.com/en/Products/Bamboo/~/link.aspx?_id=1D60C6DDB67A4A01B79A81AD39C3EDB5&_z=z

[15] http://www.wacom.com/en/Products/Bamboo/~/link.aspx?_id=877DF9CCA96144F0BA0332CED36EF468&_z=z

[16] "Wacom Bamboo Pen & Touch Graphics Tablet Peripheral review - Trusted Reviews" (http://www.trustedreviews.com/Wacom-Bamboo-Pen---Touch-Graphics-Tablet_Peripheral_review). Trusted Reviews. .

[17] posted on August 30th, 2011 (2011-08-30). "Wacom's Inkling Captures What You Draw On Paper Digitally (Amazing Video)" (http://techcrunch.com/2011/08/30/wacom-inkling/). TechCrunch. . Retrieved 2011-10-06.

[18] "Wacom eStore > Inkling > Inkling > Inkling" (http://eu.shop.wacom.eu/Inkling/Inkling?c=69138). Eu.shop.wacom.eu. . Retrieved 2011-10-06.

[19] "Wacom Forum - Wacom Europe GmbH • View topic - Intuos4 nibs flattening quickly" (http://forum.wacom.eu/viewtopic.php?f=2&t=1438). Forum.wacom.eu. . Retrieved 2010-07-10.

[20] "ImagineFX - Wacom nibs" (http://community.imaginefx.com/forums/thread/283298.aspx). Community.imaginefx.com. . Retrieved 2010-07-10.

[21] "Sijun Forums :: View topic - Wacom Tablets" (http://forums.sijun.com/viewtopic.php?t=42287&sid=38fc86b8bc5f5881a0020f4702db30f0). Forums.sijun.com. . Retrieved 2010-07-10.

[22] "deviantART Forum: wacom bamboo pen nibs" (http://forum.deviantart.com/hardware/general/1041727/?offset=30#comments). Forum.deviantart.com. . Retrieved 2010-07-10.

[23] "How long should a Wacom Intuos3 pen nib last? - GFXartist.com - Served over 20,000,000 artworks" (http://www.gfxartist.com/community/forum/145515). GFXartist.com. . Retrieved 2010-07-10.

[24] "Wacom Bamboo Owners: How's Your Nib?" (http://graphicssoft.about.com/b/2009/02/03/wacom-bamboo-owners-hows-your-nib.htm). Graphicssoft.about.com. . Retrieved 2010-07-10.

[25] "Intuos4 vs Intuos3 nib question - ConceptArt.org Forums" (http://conceptart.org/forums/showthread.php?t=155410). Conceptart.org. . Retrieved 2010-07-10.

[26] "Wacom eStore" (http://eu.shop.wacom.eu/Spare-parts-and-accessories/Intuos4/Surface-overlay-for-Intuos4-L-PTK-840?c=68865). .

[27] http://linuxwacom.sourceforge.net/

[28] http://www.thinkyhead.com/tabletmagic/

[29] http://sourceforge.net/projects/osx-pl2303/

[30] "United States (expired) Patent US4878553 describing the technology" (http://www.freepatentsonline.com/4878553.pdf) (pdf). Freepatentsonline.com. . Retrieved 2011-10-06.

External links

- Wacom (http://www.wacom.com/)
- Bamboo Apps (http://www.bamboo-apps.com/)
- Wacom Components Page (http://www.wacom-components.com/)
- Wacom Pen Ergonomics (http://www.wacom-asia.com/technical/ergonomics/ergonomics.html)
- Linux Wacom Project (http://linuxwacom.sourceforge.net/)

Soft_key

A **soft key** is a button flexibly programmable to invoke any of a number of functions rather than being associated with a single fixed function or a fixed set of functions. A soft key is often located alongside a display device of a portable device such as a cellular phone, where the button invokes a function described by the text at that moment shown adjacent to the button on the display. However, soft keys are also found away from the display device, for example on the sides of cellular phones, where they are typically programmed to invoke functions such as PTT, memo, or volume control. Soft keys are generally found on keyboards (e.g. the F keys), cellular phones, Automated Teller Machines, Primary Flight and Multi-Function Displays, although they are also found elsewhere. Hard key is the contrary term, which means a hard-coded key such as a number key pad or the Send/End key of a mobile phone. Depending on the development company, cellular phone soft keys are often called F1(Left), F2(Right).

Mobile phone

A typical mobile phone has soft keys located at left(LSK), right(RSK) and center(CSK). Depending on the modality of the application, various functions can be mapped onto it. It can also bring up multiple functions listed on a pop-up expanded menu. Usually the prompt text on the display for the soft key is not allowed to be truncated or omitted with ellipsis. The soft key itself is usually not printed with a functional icon or text, but is often marked with a dot or short bar. An example of a phone which uses Soft key is the Nokia 8810.

See also

- Screen-labeled function keys

References

- Kiljander, Harri (2004) "Evolution and usability of mobile phone interaction styles" Helsinki University of Technology, dissertation
- Lindholm, Christian. Keinonen, Turkka, Kiljander, Harri (2003) "Mobile usability : how Nokia changed the face of the mobile phone" New York : McGraw-Hill

Article Sources and Contributors

Samsung_Galaxy_Note *Source*: http://en.wikipedia.org/w/index.php?title=Samsung_Galaxy_Note *Contributors*: Abdul raja, Acalamari, Alexh7980, Anirudh12345, Benlisquare, Blackegi1, Bo yaser, C628, Cncxbox, Daniel.Cardenas, Dany Jansene, Desplow, Dickmojo, Dqeswn, Elk Salmon, Fluffylouis, Ganeshk, Icecreampop3, Jack545, Jim.henderson, Kalpesh.v.mistry, Macedoniarulez, Milan Keršláger, Moogwrench, Piotrek54321, Raysonho, Raúl González Santos, Rusirascal, Rwxrwxrwx, S4K, SF007, Samatarou, SempaiX, Slims1179, Speculatrix, Stevage, Suvodeep12, Tt100, Ykhwong, 51 anonymous edits

Android_(operating_system) *Source*: http://en.wikipedia.org/w/index.php?title=Android_%28operating_system%29 *Contributors*: 336, 777sms, 9th jinchuriki, A bit iffy, A.sutton, A520, A5b, A665321, Abhishek191288, Abrahami, Aceleo, Acery, Adambiswanger1, Adamjacobd, Adamwatters, Adi19956, Adileader, AdjustablePliers, Adm.Wiggin, Afriza, Aftekology, Aftershave, Agentlame, Akbarzpro, Akshayjain123, AladdinSE, Alboran, Alejo2083, Alex, AlexKucherenko, AlexMS, Alexey Izbyshev, Alexius08, Alfstar1997, Ali'i, Aliendude5300, Alisha.4m, AlistairMcMillan, Allen Moore, Allstarecho, Alunphillips, Alvestrand, Amarenderjannu, Amatulic, Ambictus, Amckern, Ameliorate!, Ancheta Wis, Andareed, Anderssl, Andreas Bischoff, Andreas Carter, Andrejavus, Andresfi, Andrew Delong, AndrewHowse, Andrewkantor, Androidliscence, Androidmids, Andyjsmith, Anindya Bakshi, Ankitasdeveloper, AnonMoos, Anoopan, Anoopmichael, Anthonynon, Antnee, Aoeuser, Apobilgin, Aquarat, Aradius, Arancı, Arc Orion, Arcanis, Archangelsk, ArgetlahmSource, Arghya139, Arichnad, Arjun G. Menon, Arthas01, Aryamanjain, Aryndar, Ash Crow, Ashwin18, Astonmartini, Asymmetric, Athzai Khaine, Atlantia, AtteL, Attilios, Audi152, Audriusa, Automate, AwamerT, Axl, Ayd00, Ayd000, Ayd86, Aydceri, Aydcery00, Aydcery86, Aydchery00, Aydin00, BY.Apps, Bagelfat101010, Bahua, Bangbang.S, Banksbr2, Barek, Barte, Bayonetblaha, Bbaumer, Bbisgard, Bdesham, Bedna, Beland, Bender235, Benjaminb, Benlisquare, Berelv, Betmenko, Bevo, Bhny, Bilbo571, Binarybits, Bios Element, Blackfireshocker, Blahbabe61, Blindwaves, Blogfactor, Blowdart, Bmwtroll, Bomazi, Bonadea, Bosqueschool, Bovineone, Bpave777, Brandorr, Brianreading, Brianski, Briantist, Bryan.burgers, BucsWeb, Bungalowbill, Bweono, C628, CCalo, CJLL Wright, CJMiller, CPGustafson511, CRGreathouse, Cacophony, Caltas, Can't sleep, clown will eat me, CaribDigita, Carlton.northern, Carrlos, CaseyBorders, CastAStone, Causa sui, Cbmaster, Cbr1000f, Cburnett, Ch Th Jo, Chainz, Chancer1001, ChaosData, Chardot, Charles.h.white@gmail.com, CharlesC, Cheekeong123, Cherie327, Cherkash, Chezi-Schlaff, Chiles Malesters, Chillpenguin, Chillum, Chirags, Chr1syr, Chris Bainbridge, Chris the speller, Chris.p.wu, ChrisHeller, Chrismiceli, Chrispilot2293, Christian75, Ciphergoth, Ciphers, ClaudeReigns, Clovis Sangrail, Coffee, CommonsDelinker, ComputerGeek706, Conan, Conti, Conzorz, Coolbho3000, Coolstoryhansel, CoordinateFreak, Corevette, Count Chockula, Cpl Syx, CrabbyPatrick, Craigbarnes85, Craigbrass, Cresdajv, Crysb, Csrempert, Ctjf83, Cybercobra, Cyrotux, D-Notice, D20sheets, DGMDGM, DHN, DStoykov, Daabomb, Daev, Dale Arnett, Dancter, Dandyandroid, Danger, Daniel.Cardenas, DanielPharos, DarTar, Dark-Fire, Darkspy945, DarrenW, Darrenm540, David Edgar, David Woodward, Davidhorman, Dawnseeker2000, Dbachmann, Dcxf, Deattitude, Degorr, Demysc, Denniss, Dennisthe2, Desbest, Dhaluza, Diamondland, Diannaa, Diblidabliduu, Diego Moya, Diego.viola, Digana, Digilee, Dingar, Dismas, Dj.cowan, Dlrohrer2003, Dmit, DmitryKo, Docu, Dontmitchell, Douglaswth, Download, Dpupkov, Dra.vladvamp, Dragon 280, Drbreznjev, Dreaded Walrus, Drogonov, Drrll, Dsh13, Dsrivallabha, DudeThinking, Dudyk, Dueynz, Dvyjones, Dziedrius, E258, E2eamon, EEMIV, ESkog, Eaefremov, Eagle-slayor, Eapache, Ebe123, Echeese, Eciepecie, Ed Burnette, Ed Poor, EdBever, Edoe, Edward, Ej159, Ekerazha, Electron9, Elektron, Eleman, Elronxenu, Emurphy42, EngineerFromVega, Enigmaticland, Ennustaja, EonOmega, Erc, Eric 324, ErkinBatu, Esebi95, Espertus, Essayemyoung4009, Estemi, Ettrig, EugeneKay, Eugrus, Exien, Explorer25, Faddykeyboard, Fanatic.manav, Fangfufu, Fastily, FatalError, Fattmann, Fbtjock, Feedmecereal, Ferengi, Ffinder, Filmore, Finalius, Firefoxian, Fish and karate, Flatterworld, Fleminra, FlieGerFaUstMe262, Flintb, Flohack, Fluffylouis, Found5dollar, Fran Rogers, Frap, FredTubale, Freddicus, Free French, Friginator, Fryn, Fsamuels, Furrykef, Fxhomie, Gabriel A. Zorrilla, Gainesk, Gaius Cornelius, Galaxytab, Gallagher783, Gary King, Gautamkishore, Gboxdance, Gdm, Geary, Gegorg, Gerhman, Gh5046, Ghepeu, Ghettoblaster, Ghost650, Glany222, GlasGhost, GlassCobra, Glen 3B, Goa103, Gogo Dodo, Gogoloid, GoingBatty, Gokberks, GoldKanga, Golftheman, Good Olfactory, Googlemobileplatform, Googlesubculture, Gordon Ecker, GorillaWarfare, Gouranga Gupta, GraemeL, Grafen, Graft, Graig123, Grandscribe, GreenpeaceUbuntuMan, Gregconquest, Gregory Heffley, Gronky, Gsarwa, Gscshoyru, Gsonnenf, Gu1dry, Gudeldar, Gugu2903, Guitarguy99081, Gurch, Guru4321, Guyjohnston, Guzzyron, Gyro Copter, H4lfN3ls0n, HJ Mitchell, Haakon, Hacheema, Haggisfarm, Hammersoft, Hanifbbz, HardCorwen, Harizotoh9, Harp, Hcaandersen, Headinthedoor, Hedge777, Henriok, Henry W. Schmitt, Herakleitoszefesu, Hervegirod, Hgb asicwizard, Hockeyc, Hominid, Hoss789, HotXRock, Hotcrocodile, Howlingmadhowie, Htchien, Htinlinn90, Hu12, Hucz, Hughcharlesparker, Huku-chan, Hutchinsonam, Hydrox, Hymek, I Feel Tired, I, Podius, I5bala, IBoy2G, IGEL, IRWolfie-, Ian1337, IceHunter, Icep, Icydesign, Ijon, Iknowyourider, IlPisano, Illegal Operation, Imagine Reason, Immunmotbluescreen, Imroy, Indianstar, InternetMeme, Invenio, InverseHypercube, InvertedPendulum, Ionistii, Iridescent, Irishguy, Irislia, Ironmagma, Isaacwaller, Island Monkey, Iuhkjhk87y678, Ivant, Ivario, J.delanoy, JAMJAM1666, JEN9841, JHunterJ, JLMCGE01, Jack007, Jacob Poon, JacobSheehy, Jacobmathias, Jadden14, Jaizovic, Jamadagni, James Foster, JamesBWatson, JamesNBarnes, Jamgraham88, Jamougha, JaredMT, Jasper Deng, JavierCane, JavierMC, Jb0807, Jboyens, Jbreckenridge, Jcogbil, Jdthood, JeR, Jeff G., Jeffq, Jeffrey Sharkey, Jenova20, Jerebin, Jeromeds99, Jerrinsg, Jerryobject, Jesant13, Jessica23, Jhonnyx1000, Jiess, Jim1138, Jimmin, Jimthing, Jimv1983, Jinmyo, Jmcdon10, Jmecimore, Joconnor, Joemalt1832, Johantheghost, John Ericson, JohnSawyer, Johnconorryan, Johndburger, Jokonek, Jonabbey, Jonathan-Morris711, JonathonSimister, Jonkerz, Jontintinjordan, Jopo, Jorge Stolfi, Joriki, JosJuice, Josh.e.stroud, Jpvinall, Jreferee, Jrishel, Jtangsw, Jtfcobra, Jubeidono, Julesd, Jurisnipper, Jusses2, Jvosika, Jwkilgore, KAMiKAZOW, KDesk, Kaicarver, Kaisersushi, Karam.Anthony.K, Karthickmad, Katherine, Katoh, Kawasakik, KayoWikiP069, Kenny Strawn, Kentyman, Kevin James Field, Kevthegreat55, Kforeman1, Khalid hassani, Khanayub1986, Khr0n0s, Kiddington, Kien64, KimDabelsteinPetersen, Kingdowney, Kingpin13, Kiore, Kitsunegami, Kkm010, Klingoncowboy4, Kmdowns, Knud Winckelmann, Koavf, Kokken Tor, Komitsuki, Kontar, Korkut00, Korkut000, Kozuch, Kraftlos, Krazywrath, Krc1185, Kris cs1, Kronox android, Ksyrie, Kungming2, Kushal one, Kylelnny, Kylesamani, LSUniverse, LafinJack, LancerEvolution :, Lanilsson, Larrymcp, LarsHolmberg, LarsPensjo, Lbstone27, Legoboy920, Lenar, Lesmin, Lester, Lethe, Leuko, Levineps, Lexischemen, Lfcohen, Libcub, LightSpeed3, Lightenoughtotravel, Lightmouse, Lindamilton, Lindberg, Lkt1126, Llancast, Llewelyn MT, Logan, Logical Cowboy, Lokpest, LookingGlass, Lopifalko, LorenzoB, Lotje, Lovetinkle, Lucas.Yamanishi, Luckerr, Lun Esex, M0sia1, MER-C, MZMcBride, Mac, Macungie, Magioladitis, Mahanga, Male1979, Manop, Mantrik00, Mappum, Marc Lacoste, Marcus Brute, Marcus Qwertyus, Marcus2020, Mardus, Marek69, Mark Renier, Mark0528, Markmcwiggins, Marko Gargenta, Markpb91, Marksbark, Marqueed, Martin.komunide.com, Mastrsushi, Materialscientist, Mathewsherdil, Matt Darby, Matthew0028, MatthewBurton, MattieTK, Mattkap, Mattkap2, Maulikdave05, Maurice Carbonaro, Mauripop, Maxdeutc, Maximus06, Maxí, McGeddon, Mcld, Mdikici, Meepzip, Meersmaj, Melab-1, Melmann, Mendaliv, Mentifisto, Mephistophelian, Messiisking, MetaManFromTomorrow, Mewtu, Mharen, Michaelplourde66, Midgetman433, Mikael Häggström, Mike Rosoft, Mike.lifeguard, Milan Keršláger, Mild Bill Hiccup, Millstream3, Miltonhowe, Mimihitam, Mindmatrix, Minterior, Mirabilos, Miserlou, Mistral Mktg, Mistral Solutions, Mkouklis, Mobilepush, Modamoda, Mohanpram, Moneytoo, Moocha, Mordka, MoreNet, MoreThings, Mortense, Mr. Met 13, MrGALL, MrOllie, Ms2ger, Muelaner, Mugsywwiii, Mugunth Kumar, Mutchy126, Myscrnnm, N2e, N5iln, NYKevin, Naddy, Nagy Dániel, Nagytam, Nahado, Namures, Nantasatria, Nathanloop, NeMeSiS, Nealmcb, NeilN, Neinsun, NetHunter, Newsoxy, NexuSix, Nexus26, Nicholas Love, Nick Garvey, Nico357, Nightscream, NiklasBr, Nikpapag, Ninly, Njonji, Nodekeeper, Nogburt, Noozgroop, Norm mit, Now wiki, Nuclearmoose, Nuujinn, Nyco, Ofennell, Ohaaron, Ohnoitsjamie, OlavN, Old Number7, Oldmokmok, Oleg Alexandrov, Oli Filth, Omshivaprakash, Orange Suede Sofa, Originalwana, OsamaK, OspreyPL, OwenBlacker, Oxwil, P.Shack, P2jones, PILZI, Papatenor, Pascal.Honore, Patrick, Paulmlieberman, Paulscrawl, Pdfpdf, Peter712, Peterkagey, Pgan002, Phalinshah, Phatom87, Philip Trueman, PhosphoricX, Phy1729, Piast93, PieterDeBruijn, Pilif12p, Pinball22, Pinecar, Pjedicke, Pkkasu, Plankhead, PlantRunner, Plarem, Plop, Pluma, Pmod, Pmyteh, Pokstad, Pol098, PolarYukon, Pomegranate, Poooooooooo123, Potentials, PriceChild, Prius 2, Privateboz, Procedure, Prolog, Prosfilaes, Prototypecreative, Pryanni, Psantora, Pseudomonas, Pvanderlee, Pwnage97, Quarkgluonsoup, Quartermaster, Quebec99, Quoth, Qwyrxian, RScheiber, Raburton, Rachel263, Raghualluri, Rajanbalana, Rajeshsweb, Ral725, Ralfsmouse, RameshaLB, Ramonrabello, Random name, Randomname66, Rapjul, Rapomon, Ravipokemon, Raysonho, Rborghese, Rchandra, RcketScientist, Reaper Eternal, RedHillian, Redekopmark, Reebsauce, Reedy, RegenerateThis, RenniePet, Res2216firestar, Resplendent, Rich Farmbrough, Richard Arthur Norton (1958-), Richi, Richiekim, Riffic, Rigelt, Riki, Ringbang, RingtailedFox, Rjwilmsi, Rmanke, Robbrown, Robert Moyse, RobertMfromLI, Roberth Edberg, Roberto.larcher, Robferrer, Robzz, Rocboronat, RockMFR, Rockysmile11, Rod92p, Rodeosmurf, Roguegeek, Roif456, Ronnies1312, RotaryAce, RoyBoy, Royce, Rprpr, Rugops, Runtime, Rush2009, Ryan8374, RyanQuinlan, RzR, Rzęsor, S1encing, SF007, Sachinchavan.in, Sagarwal1981, Saifuddinap, Sailsbystars, Sainath468, Salamurai, Salazasu, Salvio giuliano, SamJohnston, Samdman95, Samkass, Samuh, Sandstein, SarekOfVulcan, Sasank, Sayden, Sbmeirow, Scampy11, Sciencewatcher, Scientus, Scl98029, ScottyWZ, Seanjacksontc, Searchmaven, Secretlondon, SephirothXIIIX, Seven.cardwell, SeyedKevin, Sfm 7, Shachar, Shadez08, Shadowjams, ShakataGaNai, Shaolinx, SharkD, Shaswat Narendra, Shevett, SidP, Sierra1bravo, Sigma2488, Sijil cv, Silvio Marano, Simonrleung, Simple Bob, Sjl0523, Skier Dude, Skierpage, Skudo900630, Slakr, Slatedorg, Sleepy Sentry, Sligocki, Small.is.powerful, Smashville, Smitty, Smyth, Snakeskincowboy, SoWhy, Socialmaven1, Solinym, Solipsys, Solomon Douglas, Soma6, Some jerk on the Internet, Someguy1221, Spaghetti64, Speculatrix, Spiel, Sreyan, Sriram sh, Staka, StaticGull, Steel, Stefan, Stephenb, Stephenwanjau, Steve03Mills, Steve1428, Stevedel7, Steveklein, Steven Walling, Stevenbz9, Stevenmitchell, Stevenwagner, StewieK, Strcat, Subbu, Suckystraw, Suction Man, SudoGhost, Sukael, Sunnypsyop, Sunray, Suzals3, SvGeloven, Svetovid, Svick, Swampyank, Sygmoral, Sylvainchevalierfu, Syndicate, Syp, T-Rex84, TMO KOTOR, TX159, Tagrb03f, Tahitiville, Taras, TarzanJr, TastyCakes, Tavilis, Tbhotch, Tbird20d, Tcncv, Tedder, Tedickey, Teeks99, Teleprinter Sleuth, Teles, Tgeairn, The Anome, The Letter J, The359, TheEditrix2, TheTechFan, TheWishy, Thealexweb, Theanphibian, Theartfullodger, Thecurran91, Themfromspace, Thesamami, Thingg, ThomasWilson2, Thorwald, Thumperward, Thunder Wolf, Thunderbird8, Thüringer, Tide rolls, Timeshifter, TimmmmCam, Timneu22, TobiasPersson, Tobziez, Tomchen1989, Tomlzz1, Tomself1, Tondi5, Tony Sidaway, TonyW, TorQue Astur, Torqueing, Traal, Tracer9999, TrbleClef, Trebek Skates, Trefork, Tri400, Troed, Trusilver, Tsriopensourceblueprints, Tuxcantfly, Tweisbach, Txaggiemichael, U5K0, UKER, UU, Uirauna, Ujimatcha, Ulric1313, Unamed102, Unknownwarrior33, Urashimataro, Urfavoritemija, User931, Usmanahmed25, Vadmium, Van helsing, Varlaam, Vcelloho, Venona, Victorpardosi, Vincenzo.romano, Voidvector, VoluntarySlave, Vrenator, Vujke, WakiMiko, Walter Görlitz, Walterlmitchell3, Waltonkbbl, Watchcars, Wbison3, Wednesday Next, Wello95, Werbej, Werdna, Wertydm, Wesleyarchbell, WhatMeWork, Whatiknow, Wickedjacob, Wifuk, WikiLaurent, Wikigod, Wikipedian2009, Wild mine, William Leadford, Williameis, Windofkeltia, Wintermute115, Wizardist, Wknight94, Wlindley, Woohookitty, Woolfy123, Writermonique, Wtmitchell, Wwoods, XJamRastafire, Xavierorr, Xcrivener, XdaLive, Xhienne, Xiutwel-0003, Xnamkcor, Xomm, Xrobau, Xsspider, Xx3nvyxx, Y2kcrazyjoker4, YlCbaby, Yadavjpr, Yahia.barie, Yamla, YasharF, Yellowdesk, Yiosie2356, Yngvarr, Yousou, Yug, Yuriybrisk, ZacBowling, ZamorakO o, Zaratoustra, Zbutler7, Zeldex, Zero sharp, ZimZalaBim, Zipcodeman, ZirconiumTwice, Zorak950, Zouzzou, Zundark, Zunmun, ^musaz, Ævar Arnfjörð Bjarmason, Δ, Սահակ, , , 2352 anonymous edits

Smartphone *Source*: http://en.wikipedia.org/w/index.php?title=Smartphone *Contributors*: 16@r, 3Coins, A Man In Black, A123a10, A8UDI, AKA MBG, ALSLUG, AVM, Aaditya 7, Acalamari, Acather96, Aclassifier, Acolin f, Adriatikus, Aeons, Ahoerstemeier, Aillema, Akuyume, Alansohn, Alex890, Alf Boggis, Alf.laylah.wa.laylah, AlistairMcMillan, Allen3, Amarendra.Avinash, AndrewSpec, Andries, Andros 1337, Angela, AniRaptor2001, Ansible, Anupam, Arab Dynasty, Arafael, Arny, Arp120, Arunsingh16, Atama, Atenyi, Atul.ecn, Avoided, BD2412, Bahaltener, Bakshi41c, Balazer, Baronnet, Beland, Bendes, Bhny, Bibijee, Biker Biker, BillHaywood, BlkStarr, Bomazi, Bovineone, Branddobbe, Brianreading, Broccoli, CLW, CRJO-CRJO, Cactusframe, Cadiomals, Caj27, Calcwiki, CalumCook234, CanadianPenguin, Canalteen, CatherineMunro, Cellcom, ChaChaFut, Changshui88, Charivari, Chris the speller,

Ciphers, Citizensmith, Cleared as filed, Closedmouth, Clovis Sangrail, Cmf, Cmlewan, Cmp101, Codetiger, Comaleaf, CommonsDelinker, Commontimect, Cool Hand Luke, Coolaaron88, CoolingGibbon, Cornea503, Cst17, Currab, Cxtom, Cyruslei, DaisyField, Dale Arnett, Dan6hell66, Daniel.finnan, DaveChild, Dcljr, Dcxf, Ddavid2005, Deane@gooroos.com, Delphii, DerBorg, Deviceapps, Diannaa, Diego Moya, DivineAlpha, Dmarquard, Dreambringer, Drobatch, E.au, EVula, EWikist, Editor182, Editrrr, Elassint, Electronicguru1, Emma23 K, Epolk, Eraserhead1, Erc, Erianna, Eshade, EugeneZelenko, Eugenia Ioli, Evice, Evilruletheworld96, Evolutiondb, F1MotoGPWRC, FCYTravis, Farolif, Father Goose, Fcassia, Feinoha, Feydey, Fhontoy, Flowanda, Fourdee, Furrybeagle, Fuzheado, GTBacchus, Galadh, Garant^ ^, GateKeeper, Geologyguy, Georgy90, Giftlite, Giggett, Gilliam, Ginsengbomb, Giraffedata, Glacialfox, Glennwells, Gmumru, Gogo Dodo, GoingBatty, Gold Hat, Gomm, Gookey, Gordon Ecker, GraemeL, Grahamperrin, Grajales, Graywolf, Graywolfmoon1, Gregorysalt, Greswik, Groink, Gsarwa, Guy Harris, Gwernol, HDCase, Ha us 70, Haakon, Hadiceberg, Halloween.mac, Hamiltha, Hammersoft, Hardylane, Harmil, Harryzilber, Haseo9999, Hede2000, Henrik, HereToHelp, Hervegirod, Hobartimus, Hu12, Human.v2.0, Huntersquid, HuskyMoon, HybridBoy, Hydrogen Iodide, Hydrox, Hypnotist uk, I already forgot, I, Podius, Ianb, Ianboudreault, Ianereed, Icedwater, Ignis Fatuus, Illegal Operation, IllyriaO5, Im Buff, Immunmotbluescreen, Imroy, Imsilly123, In2thats12, Indefatigable, Interframe, InternetMeme, Intershark, Ipsign, Irky, Ishdarian, ItsZippy, J, J.delanoy, JCDenton2052, JCrue, JNW, Jagdterrier, JamesWeb, JamieS93, Jantangring, Jeffq, Jekyllhide, Jerome Charles Potts, Jerryobject, Jfdwolff, Jim.henderson, Jimmy Bergmark, JoeSmack, John of Reading, JonHarder, Jonabbey, Jondel, JonoP, Joseph Solis in Australia, Josh Parris, Jostake, Jsherwood0, Jsribeiro, Juan M. Gonzalez, Jusdafax, Just another editor, Justin Mauger, Jwojdylo, Jæs, Kajowi, Kariteh, KarmaGeddon, Katieh5584, Kbdank71, Kcomstock, Kellen`, KelleyCook, Kentynet, Kevin Dorner, Kingpin13, Kinst, Klokbaske, KnowBuddy, Koavf, Kozuch, Kraftlos, Kresp0, Kudpung, KungFuMonkey, Kuru, Kwaichi, LaVieEntiere, Lai888, Lambtron, Larkhill97, Laurasmith70308, Lazulilasher, LeilaniLad, Lenin1991, Lester, Leszek Jańczuk, Leujohn, Level plus, Lewispb, Lifes g00d 561, Lilac Soul, Limequat, Little Mountain 5, Little Professor, Logan, Loser997, Lotje, LrdChaos, Lukan4ica, Lun Esex, Lun4tic, MER-C, MMuzammils, Mac, Mac John Concord, Magnus.de, Malik Shabazz, MangeO1, Maniacgeorge, Manop, Manway, Marcisjustice, Marco.difresco, Marcus Qwertyus, Mark Kim, Mark91, Marksza, Marrowmonkey, Martarius, Masgatotkaca, Mathiastck, Maury Markowitz, Maxí, Mcduck, Mckote, Mdrejhon, Mdwh, Meehawl, Meelar, Memaster3, Mephistophelian, MetaManFromTomorrow, Miaow Miaow, Mikael Häggström, Mike Dillon, Mike Linksvayer, Mikehelms, Mild Bill Hiccup, Minghong, Miquonranger03, Mix Bouda-Lycaon, Mlg07e, Moberg, ModWilliam, Monaarora84, Mononomic, Morphh, Mortense, Mosmof, Mr. Strong Bad, MrOllie, Mvjs, My76Strat, MySchizoBuddy, MyronAub, Myscrnnm, Nakakapagpabagabag, Nathan94124, Navacell, Nazrich, NeOFreedom, Nealmcb, Neo Ogilvy, Neoarchon, Neoguy999115, Nerdeff, Netrapt, Nick Number, Nickolai 420, Night Gyr, Nightscream, Nitro.ajb, Nixdorf, Nneonneo, Nopetro, Nurg, Obey, Ohnoitsjamie, Oknazevad, OlavN, Old port, Oli Filth, Omegatron, Oontay, OpenToppedBus, P.Marlow, PS., PTSE, Parintachin, Patrick, Pats1, Patvac-chs, Paul 1953, Pbrown111, Pdahomepage, Petershank, Peyre, Phatalflaw, Phatom87, Phearson, Philip Trueman, Piano non troppo, Pingveno, Piotrek54321, Pluma, Pmarshal, Pmlineditor, Pol098, Pol430, Poor Yorick, Posix memalign, Prathameshsasane, Prillen, Pvanheus, Pyfan, Quebec99, Querencia, R'n'B, RA0808, Radiier, Radon210, Rafael.sp, Ramu50, RedWolf, Reliablesoft, Repetition, Res2216firestar, ReverseEngineered, Reyk, Rgreed, Rhobite, Riadlem, Rich Farmbrough, Richardguk, Richiekim, Riki, Rollins83, Ron2, Ronz, Ruchir257, Rursus, SF007, SPQRobin, Sadegh87, Sainath468, Samad120, Samwb123, Sandstein, Sapibobo, Sayid98, Schrödinger's Cake, ScottyBerg, Scrtcwlvl, Sdfisher, Sdrazfar, Sean13zz, Sebastian Mandrean, Sebindcruz, Secretsorry, Serg3d2, Sfacets, Sfan00 IMG, Sfsmartp, Shalroth, Shawnc, Sheehan, Signalhead, Siliconov, Singerdg1, SirJibby, Skatebiker, Skintigh, Skittle, Slitchfield, Smart1954, Smmgeek, Snigbrook, Soliloquial, SpaniardGR, Spellmaster, Starofwonder, Stephen B Streater, Stephen Turner, Stjson, SuperHamster, Surenkarapetyan, Suwatest, Swoof, TVS99, Taka76, Takamaxa, Taketa, Tangent1000, Taskinen, TastyPoutine, Tatterfly, Technopat, Tesi1700, Tfgbd, ThaWhistle, ThaddeusB, The Pink Oboe, The Thing That Should Not Be, The Wild Falcon, The-apathy, Theaveng, Thestick, Thiled, Thisma, Thumperward, Ticell, Tide rolls, Tikru8, TimSE, Tklaer, Tmuller2, Tobias Bergemann, Togopogo, Tokyogamer, Tomas.turek18, Tombomp, Tommy2010, Tommytentimes, Tony1, Toussaint, Tpbradbury, Trasz, Treekids, Tuju, UCLATre, UnrealG, Urod, Uzume, Vegaswikian, Verkinto, Verne Equinox, Versus22, Vkem, Vrenator, Waqas Hussain, Westie4321, WhisperToMe, Whkoh, WiZZLa, Wideangle, Wiki admi, Wikimhb, WikipedianYknOK, Wikipelli, Winged-stone, Wireless Buddy, Wknight94, Woohookitty, XGrape, Xajel, Xrobertcmx, Yean3d, Youxiarock, Yug, Yunshui, Zach.vega1, Zanter, ZimZalaBim, ZipoB, Zpetro, 1100 anonymous edits

Tablet_computer *Source*: http://en.wikipedia.org/w/index.php?title=Tablet_computer *Contributors*: 842U, A Quest For Knowledge, AV3000, AakashTablet, Ace of Spades, AgadaUrbanit, Ahunt, Aizuku, Alisha.4m, Alvestrand, Ancheta Wis, Anders Feder, AndrewHowse, Andries, ArtsMusicFilm, BD2412, BKfi, Bender235, Bhny, Bidofthis, Bigredsky, Bikepunk2, Bob bobato, Camilo Sanchez, Chrissb, Chuck369, Coercorash, CommonsDelinker, Crimsonmargarine, Crysb, Cvbommel, D.M. from Ukraine, DMY, DVdm, Daniel.Cardenas, Dawnseeker2000, Dbachmann, DenisRS, Dialectric, Diamondland, Diego Moya, Dondegroovily, Drnick2, Edcolins, Editor2020, Edkollin, Elandy2009, Elizium23, Eog1916, Eraserhead1, ErkinBatu, Esebi95, EuTuga, Evan-Amos, Exlixe, Falcon9x5, Fieldafar, Fikri RA, Filing Flunky, Fishnet37222, Forresttsao, GSL-Nathan, Glenn, GreenZeb, Gsarwa, Hakimio, Heavyrain2408, Helmuthva, HiddenIP, Hydriz, Ionutpopa, Ipadtablet, Itsmine, IvanLanin, JaGa, Javabyte, Jeffnailen, Jerryobject, Jgera5, Jim.henderson, Jinnai, JohnSawyer, JonathanDP81, Joy, Jpod2009, KKoolstra, Karlson2k, Karlww, Ketil, Kevdave, Krauss, Kyng, Leofranco2000, Lexischemen, LilHelpa, Lore1996, Lppa, Lun Esex, MER-C, Macromediax, Mahjongg, Marcos, Marcus Qwertyus, Materialscientist, Mckinley99, Merbabu, MetaManFromTomorrow, Michael Minh, Mikitei, Mr Stephen, Nateclev, Nesvarbu02, New Thought, Newone, Nijusby, Noozgroop, Ohconfucius, OlavN, OspreyPL, PeregrineAY, Piotrus, Pnm, Proskillr, Red Act, Ric36, Ricvelozo, Ronz, Rostz, SF007, SamJohnston, Samarth.karwal, SarekOfVulcan, Sbmeirow, ScottSteiner, Scottywong, Shahimbaker, Shrish, Simonrleung, Socialmaven1, Soundvisions1, SpareHeadOne, Squids and Chips, Suso, Swarm, Szente, TMV943, Taka76, Tankwan, Taxman, The Pikachu Who Dared, The Thing That Should Not Be, Tide rolls, Tomasohara, Tony Sidaway, Toussaint, Transmanche, Trevj, UCDS, Vanuan, Vrenator, Vyx, WOSlinker, Wbrito, Weylin.piegorsch, WilliamBrain, Woohookitty, WorldBrains, XJamRastafire, Xcvista, Yosh3000, Zc456, Ö1mageri, 201 anonymous edits

Stylus_(computing) *Source*: http://en.wikipedia.org/w/index.php?title=Stylus_%28computing%29 *Contributors*: Ajbp, Bender235, DGG, GrahamColm, Juan M. Gonzalez, Lester, Mootros, Pmarshal, Rjwilmsi, RobertHelm, West.andrew.g, Woohookitty, 11 anonymous edits

Internationale_Funkausstellung_Berlin *Source*: http://en.wikipedia.org/w/index.php?title=Internationale_Funkausstellung_Berlin *Contributors*: Alexius08, Ark25, Bethan 182, Charbax, David Haslam, Denhis, Denniss, DerGolgo, Discospinster, Hébus, Imroy, IndianGeneralist, Lars T., Lotje, Malachias111, Meisterkoch, Michaelmas1957, Mikeo1938, Msubasi, Nihonjoe, PigFlu Oink, Prunk, Pwjb, Redevent, Renredam, Repetition, Ricardo Santiago, Richmd, Schily, Search4Lancer, TCY, Tony1, Trident13, Twang, 33 anonymous edits

AT&T *Source*: http://en.wikipedia.org/w/index.php?title=AT%26T *Contributors*: 28bytes, 5 albert square, 72Dino, 831JMO, A Man In Black, A Stop at Willoughby, A3RO, A8UDI, ACSE, AEMoreira042281, AJTWWE, Aaron7chicago, Academic Challenger, AceofdataBase, Acroterion, Acsint, AdamDeanHall, AdjustShift, Admrboltz, Afiler, AgadaUrbanit, Age Happens, Agvulpine, Aido2002, Aj41101z, Al Lemos, Alansohn, Alerante, AlistairMcMillan, Allamericanbear, Allen3, Allstarecho, Almcshane, Almusinc, Aloughman, Alpha 4615, Alphachimera, Alphathon, Alpta, Alx xIA, Amcl, Amextex, Andareed, Andonic, Andrewlp1991, Andrewpmk, AndySimpson, Anthony Appleyard, Apeloverage, Aquemini, Are you ready for IPv6?, Arghya139, Aristophanes68, ArkansasTraveler, Armbrust, Arthur Smart, Arukun14, Aspects, Assassinoc714, Astuishin, At&t1000, At&tisterrible, Atlantacitizen, Attitude2000, Auntof6, Awsumgrrl, Awtribute, Ayengar, B64, Babbage, Bachrach44, Badagnani, Badmachine, Bakbro2, Barek, Barryob, Baseballbaker23, Bataluis, Beetstra, Before My Ken, Beland, Benjaminleebrackman, Benstown, BhamAla, Bjj07, Blackcap93, Blue520, Bluepoint951, Blurpeace, Bnitin, Bob bobato, Bobbyj1049, Bobo192, Boleyn, Bongwarrior, Boo Boo Pa Doo, Boothy443, Bortain, Brain1979, Brian0918, Brighterorange, Brownings, Bruinwolverine, Burner0718, C.Fred, CJS102793, CR85747, Calvin 1998, Calwatch, CambridgeBayWeather, Can't sleep, clown will eat me, Capricorn42, Capt. James T. Kirk, CaribDigita, Cassius1213, Casull, Catapult, Catdude, Centrx, Ceyockey, Ch Th Jo, ChadCloman, Chadlupkes, Chairlunchdinner, CharlotteWebb, Cherrydude, ChetTruthseeker, Chris the speller, Chrisbw, ChristalPalace, Chupon, Ckatz, Cla68, Clippark, Clipper471, Closaveng, Closeapple, Coffeehood, Colorado Locks, Colorvision, Commander Nemet, Comrade Tux, Conrad.pramboeck, Conti, Cool Hand Luke, Coolhandscot, Corgy.x, CorpITGuy, Cotojabee, Courcelles, Cpitty2, Cpl Syx, Crocodile Punter, Crosscountryrunnur, Crownjewel82, Ctjf83, Cumulus Clouds, Cyanolinguophile, Cybercobra, Cyrusc, D'Agosta, DARTH SIDIOUS 2, DIEXEL, DMG413, DMacks, DaDrought3, Dale Arnett, DallasOConner, Danactro, Danhash, Dante Alighieri, Dave Farquhar, Daveswagon, Davidvasta, Dawnseeker2000, Dbmandk, Dcandeto, Deathnova2, Debresser, Decora, Dedust57, DeltaQuad, Deville, Dfwcre8tive, Diasimon2003, DivineAlpha, Djtrooper96, Dknights411, Don-Don, Dongk, Down2000, DowneyOcean, Dppowell, Drbreznjev, Dsayles08, Dtobias, Dubchubbub, Dunee, Dvavasour, DylanW, E2eamon, EJDyksen, ESpublic013, Eakerr4, Edgar181, Edmccray, Edward, Edwy, Eeekster, Eenu, Ehjort, Ehurtley, Ejohnsequilar, Elkman, Elvey, Emailchrishall, EmmyMaven, Empoor, EoGuy, Epbr123, Erik16, ErikNY, Estoy Aquí, Everyking, Evil saltine, Excirial, ExplicitImplicity, FF2010, Fairlyoddparents1234, Falcon8765, Falcorian, Fctchkr, Firai, Flamurai, Flavious27, Fnlayson, FrankCostanza, Fudoreaper, FullMetal Falcon, Funnyman1234567890, Fuzzy510, G27, GT, Gai-jin, Gary King, Gba030, Gegenwind, Geni, Geopgeop, Gfoley4, Giants27, Gloriamarie, Gogo Dodo, GoingBatty, Golbez, GoldDragon, Goldfish007, Golgofrinchian, Gr1st, Greenshed, Gregspencewolf, Ground Zero, Gruze7, Guaka, Gurch, Guy Harris, HA5TY, Haakon, Hahahamuamua, Halfd, Harmil, Harryzilber, Hateless, Hawaiian717, Hbdragon88, Headbomb, Heirpixel, Henry W. Schmitt, Hiberniantears, Highlight Guy, Hotcrocodile, Hu12, Hypernick1980, Ian.thomson, Ian3055, Ice Ardor, Icebrg, Idaltu, Ief, Ignatzmice, Ikip, Iliketimmyturner, Immunize, Imzadi1979, Indguyintx, Infinity Squared, Infomonster22, Inter, Iostoleggendo, Ipatrol, Iridescent, Irishguy, Isarian, Ivanmiter, J. Nguyen, J.delanoy, J3ff, JForget, JabberWok, Jafwela, Jake Wartenberg, Jamcib, Jasonstewart11, Jauerback, Jayron32, Jdmfreak, Jebba, Jediknil, Jeff G., Jeni, Jerdog, Jessamyn, Jgunaratne, Jim.henderson, Jim1138, JimScott, JimmB, Jjacobsmeyer, Jlin, Jmet321, JoanneB, John, John K, Jolomo, Jommeke, Jonathunder, Joost Hageman, JordoCo, Josephf, Joshua Issac, Jossi, Joy, Jpgordon, Jps57, Jrtayloriv, Juliancolton, JustinTSampson, Jvcdude, Jéské Couriano, KGasso, Kab823, Kainaw, Kam300062, KansasCity, Kate, Kaybyte, Kbdank71, Kbolino, Kcasllab42, Ke5crz, Keizers, Kendrick7, KennethHan, Keyblade12344, Kingoomieiii, Kitch, Kiwi128, Kjkolb, Kkm010, Kmorozov, Kn9359, KnowledgeOfSelf, Koman90, Kozuch, Kungfuadam, Kungfuazn, Kuru, LP-mn, Lambertman, Lapin rossignol, Leasnam, Lesbianadvocate, Lexlex, Lgeorgehsv, Liam Braithwaite, Liam Harvard, Lightmouse, Ling.Nut, Ljenkins13, Lloydsd, Lmno, Lotje, Luatha, Lucasoutloud, Luna Santin, Lutin jovial, MCB, MER-C, MLRoach, Majoreditor, Maksdo, Malinaccier, Markell2010, Marlinz144, Marmusek, Martarius, Master Jay, Materialscientist, Mathiastck, Matt Deres, Maxí, McDisneysoft, Medcaramelldansen, Meghyo, Melted time, Metricopolus, Mets501, Mgoblue144, Mhking, Mhwang546434242, Miami33139, Michael Hardy, Michaeldsuarez, Michaelhangge, MikeLynch, MilborneOne, Minesweeper, Minimac, Miranda, Misternuvistor, Mithridates, Mkemper331, Mmernex, Mmsmsdighsdghiuhsd6, Mnw2000, Momo san, Monty845, Moocowsrock, Morannikka, Morio, Mr. Ash, Ms2ger, Mspraveen, Munford, Murphdoggiedog, Musical Linguist, Myleslong, Mytelecom, Mzanon, N. Macchiavelli, N328KF, NJA, NThomas, NameIsRon, Narutodude000, Naryathegreat, Nayra, NeuronExMachina, Neutralaccounting, Neverquick, NewTruck08, Ngroot, Nich01asx, NickPenguin, Nightscream, Nightstallion, Nlu, NoGringo, NoseNuggets, Nrpf22pr, Numlockf6, Nuttycoconut, Odbii, Ohconfucius, Ohnoitsjamie, Oknazevad, Oleg Alexandrov, Onorland, Orange Suede Sofa, Orphan Wiki, Otterathome, Otto ter Haar, PL290, Pablosecca, Passionless, Pauljaz, Pchov, Pdcook, Pfctdayelise, Phaedriel, Phgao, Pig de Wig, Plainsong, Po1216, Porchcorpter, Porticus, Privatechef, Prof JTH, ProhibitOnions, PureRED, Pwu2005, PyroGamer, QD4rmBAMA, Quadell, Quidam65, Qutezuce, RBBrittain, RFD, RJII, RLudens2009, RadicalBender, RadioFan, Rangoon11, Rashy27, Ravenhull, Razorflame, Realitycheckmate, Reginald Perrin, Requestion, Retro00064, RevelationDirect, Reywas92, Rfc1394, Rgoodermote, Rhobite, Rhollenton, Rich Farmbrough, Richard Arthur Norton (1958-), RichardNegro, Richiekim, Richmond96, Rick Block, RingtailedFox, RizFiz, Rjwilmsi, Roadrunner3000, Robguru, Ronhjones, Rory096, Roxxyroxursox, RoyBoy, Ruduen, Ruud Koot, RyanGerbil10, Ryank808, S51438, SCEhardt, SJP, SLEZ, SMC, SNIyer12, Sardanaphalus, SchmuckyTheCat, Schnell, Sciencecomp314, ScottSteiner, Scrutchfield, Seahorseruler, Sean, Shadowjams, Shane52sk, Shawnc, Shihtzusrool, Shirulashem, Shyguydie, SigKauffman, Signalhead, Sihfusih, Simaloko, Sk19.g, Sky Harbor, Skybon, Slakr, Smcm07, Smmgeek, Snigbrook, Soccerchickdf, Social Studiously, Sohailstyle, Solarisworld, Some Person, Spanginecr, Sparrowman980, Special Cases,

Speer320, Spenc84, Sponge1987, Spyder Monkey, Squiggleslash, Ssg2442, Stephen doctrin, Steven Zhang, Steven is dumb, Stevenmitchell, Stickguy, Straykat99, Stromcarlson, Su2000, Subman758, SunDragon34, Supercool94, Svgalbertian, Swbuehler, T-borg, T1980, THEunique, Tbhotch, TechGeek70, Techman224, Teckdiva, Tedder, Telepheedian, Telso, Tetraedycal, TexasAndroid, The Navigators, The Thing That Should Not Be, The Utahraptor, The wub, TheIguana, Themichaeljones, Thepangelinanpost, Tholme, Thomas Larsen, Thunderbrand, Tiddly Tom, Tide rolls, Tim1337, Tinton5, Tkaizan, TomDonohue.NewsVisual, Tomabuct, Tombomp, Tommy2010, Toonces69, Top-up update, Topbanana, Tpeltz, Tresiden, Trident13, Trusilver, Ttenchantr, Tubber Road, Twilight Secret Star Guy, Tylerscott, Tymur3, USA300million, Uarkrazorbacks1, Uncle Dick, UnitedStatesian, Unscented, UpstateNYer, Urbanrenewal, Uris, V8A2N6, VTEX, Vchimpanzee, Vertov76, Vespristiano, Vikramsidhu, Voice99, WCAWiki, WJetChao, Wadsworth, Ward3001, Warut, Washburnmav, Wavelength, Wayne Slam, Wd40, West.andrew.g, WhenEliMetLilly, WhisperToMe, Whoisthatclown, Whomp, Whywhenwhohow, WikiLeon, Wikifan1001, Wikiguy729, Wikipedianinthehouse, Wikipelli, Willking1979, Winbuyer, Wkdewey, Wmualumnus, Woohookitty, Wtmitchell, Wuhwuzdat, Wwwe1234, Wysprgr2005, X0X9, X570, XXJMOXx, Xde 13a, Y2kcrazyjoker4, Yamamoto Ichiro, Yellowdesk, Zachary0220, Ze0h4x, Zereshk, Zgillis, ZimZalaBim, Zmiller923, Zoicon5, ZooCrewMan, ZooFari, Zpb52, Zzuuzz, Δ, 1316 anonymous edits

Samsung_Galaxy_S_II *Source*: http://en.wikipedia.org/w/index.php?title=Samsung_Galaxy_S_II *Contributors*: 5 albert square, A520, Acalamari, Akwiki11, Allsunnydays, AlvinPing, AmandeepGrewal, Amire80, Anomie, Anonymous Coward 87, Aohus, Aqualung, Argus fin, Arjayay, Artem-S-Tashkinov, Arun2007th, Asliturk, Avinashega, AxelBoldt, BD2412, Badgernet, Badudoy, Bender235, Benlisquare, Bierfuizl, Big Bird, Bigmig2200, Bruce lee, Calvin Marquess, Choledocho, Chris the speller, Chrisvarns, ChromeFirst, Chzz, Closeratio, Cmc87, CommonsDelinker, Curly Turkey, Cybik, Dabix, Davemc50, Dcxf, DeltaSPARTAN003, Dethtrain, Dibyendu8t1, Djapor96, DmitryKo, Doodaaa, Download, Dsh13, Eelamstylez77, EngineerFromVega, Escape Orbit, Esebi95, Etoile, Eugenevdm, Excirial, Fauzan Zaid, Felycia, Filiprino, Foolip, FreakyDaGeeky14, Fresheneesz, GeekyPuppy, Georgie1311, Gnulinux, Gogo Dodo, Golem-red, Greenman, Gsarwa, Gum10000, Gumguy10000, Halsteadk, Hsan22, Ihatetoregister, Imroy, Informationrestorer, Inphrased, IrfanFaiz, JCB Digger, JERRKOWA, JSH-alive, Jacealcard, Jayathakrar, Jerome Charles Potts, Jneuenhaus, KDesk, KarmaLoop, Khanayub1986, Kim2jy, Kms, Kokken Tor, Lexusuns, MER-C, MMMaroko, Macedoniarulez, Maraist, Marc-Olivier Pagé, Materialscientist, Mateuszzz88, Matt Lib Soc, Matt Yohe, Michael73072, Mikel Ward, Milominderbinder2, Mingyishi, Mlaclom1, Mobilejoy, Moemoemork, Moocha, MrAtoz, MrLemon9876, Mro, Mtking, Mulad, Mumbaifreaks, Mushroom9, Myscrnnm, Mystvearn, N5iln, Olyus, Onewhohelps, Pawlen-T, Philg88, PhnomPencil, PieterDeBruijn, Piotrek54321, Pm4gis, Pmc, Pol098, PowerPatrick, Prankey, Privateboz, Psoreilly, Pthomas1991, Quickone, Rachelanytime, Raghavd17, Ramo1984, RaphaelJacob, Raysonho, Reinraum, Romdanen, Ron1n541, Ronnies1312, Ryuch, Rzęsor, S3bst3r, Saadixxx, Salzion, SamsungGalaxyS2, Seaphoto, Shba, Shukig, SiggyF, Silversword411, Sir Dagger133, Sixequalszero, Skatebiker, Skier Dude, SoWhy, Sokka54, Steel, Stoidjeg, Strippedman, Superweapons, Supreme Deliciousness, SwisterTwister, Syp, Tbhotch, TechGizmo, Telanis, TippTopp, Tommy, Trycatch, Userpd, ViperSnake151, Virtualerian, Vyker-temp, Warrenski, WhiteZero, Wigbold, Wiki timo, Wikiguy10000, Wjfox2005, Worrdo, Yadavjpr, YuMaNuMa, Zecharixs, Zeroverse91, Zhang.jiahe, Zouzzou, 602 anonymous edits

Samsung_Galaxy_Tab *Source*: http://en.wikipedia.org/w/index.php?title=Samsung_Galaxy_Tab *Contributors*: 5 albert square, 842U, Acalamari, Al Lemos, AlistairMcMillan, Amore Mio, Ancheta Wis, AndrewHowse, Aprav, Ark25, Arny, Arsonal, Arun2007th, Augustobafua, AxelBoldt, BD2412, Bender235, Bizzyjb, BokicaK, BuuBox, C628, Chambo622, Chowbok, Chris the speller, Cncxbox, DVdm, Dale Arnett, Davidkierz, Dawnseeker2000, Dblan010, Dennis Bratland, DerekMorr, Diamondland, Discospinster, Dushyanthan, Dznph3, Eatskippy, Edward, ElComandanteChe, Eraserhead1, Evefavretto, Evert:Meulie, Felixhonecker, Freddflynn, GMBFly98, Gabegom, Gogo Dodo, GoingBatty, Gryllida, Gu1dry, Gunnala, Halaqah, Hari82, Hbdragon88, Heavyrain2408, Hervegirod, Humblefool, ILikeEverythingApple123, InternetMeme, Intoronto1125, JNorman704, Jalexoid, Jim.henderson, Jjlome, KDesk, Karaff, KeiLOMs4, KeloGitiM2, Kenfied, Kesmet, Lamatrice, Lawrencekhoo, Lealvf, Lightmouse, Lm 997, Lmpandey, LostAlone, Lucky widianto, MER-C, Maddox, MadnessInside, Maru0720, Mdwh, Mernen, Michaelmas1957, Mkhydra, Mobile Smug, Mobilejoy, Modamoda, Mortense, Msousasj, NawlinWiki, Nick Number, Nimrod8, Oettam, Ohconfucius, Old Moonraker, Paulikxp, Piculo, Pol098, Pragvansh, Pratyeka, Qwavel, Raghith, Ral725, Ramaksoud2000, Ravin' Ray, Rjwilmsi, Robert The Rebuilder, Roy royles, Rrburke, Ryan31S, Sacreddeamon, Sargund, Sefford, Sgoh3781, ShelfSkewed, Shorty23sin, Soc88, Sokka54, Some jerk on the Internet, Sp33dyphil, Spacyzuma, Statisticsguy, Sunmylondon10, TMV943, Tallungs, TechGeek70, Tesi1700, TheCosmicFrog, Thedarkestshadow, Themfromspace, This is my unique user name, Timothylord, TippTopp, Tony Sidaway, Treekids, Trusilver, Virginiajim, WOSlinker, Wasell, Wikipedia XP, Winne, Woohookitty, Wsbsteven, YardOikeM7, Yosri, Yuriybrisk, Zephyrizm, Zyboris, 321 anonymous edits

Mobile_phone_form_factors *Source*: http://en.wikipedia.org/w/index.php?title=Mobile_phone_form_factors *Contributors*: -Edwin-, Allo002, AniRaptor2001, Blackboxxx, Carterman13, Dale Arnett, Df7102, Editor182, Evosoho, Fangfufu, Haikallp, Hammersoft, Imroy, Jason24589, Jim.henderson, Marcus Qwertyus, Martarius, Mathiastck, Mattgirling, Mdwh, Mtking, ONEder Boy, Pequod76, Pnm, Priveledge, RW Marloe, Red dwarf, Roif456, SF007, SchuminWeb, Scientific29, Skarkkai, Taylorluker, Toussaint, Ubcule, WikHead, Wiknerd, YUL89YYZ, Zzyzx11, 59 anonymous edits

Subscriber_identity_module *Source*: http://en.wikipedia.org/w/index.php?title=Subscriber_identity_module *Contributors*: 001.keshav, 16@r, A3RO, A5b, Abdull, Aido2002, Ale jrb, Aled D, Alex.tan, Alexburke, Amicaveritas, Andrewferrier, Andros 1337, Anthony Ivanoff, Arnel enero, Asdino, Ashkhan, Ashwin, Askari Mark, Audin, Audriusa, AxG, Bahram.zahir, Baloo rch, Bart.vanassche, Bencherlite, BigHairRef, Bogybogy, Bombay01, Bouvierjr, Bratch, Brighterorange, Bsoft, Bubbachuck, Bwpach, C Ruth, Cap'n Refsmmat, CaribDigita, Chrislk02, ChuckBiggs2, Ckatz, Clawed, ClementSeveillac, CommonsDelinker, DBlomgren, DWaterson, Dabombazzz, Daemondust, Dale Arnett, DanieLo.jenkins, Danno uk, Darac, David Traynor, DavidRCrowe, DerHexer, Dirkbb, Dkastner, DmitTrix, Dmitrytorba, DocWatson42, Dodger, Dreadstar, Dtcdthingy, Dtobias, Duckbill, Duncan, Ed g2s, Edin1, Egil, Electron9, Elv2003, EnTheMohammad, Etmjang, Excirial, Fangfufu, Foofy, Foosterhoff, Frap, Gadfium, Gamelore, Gary King, Geoff Plourde, Georgy90, Ghettoblaster, Glane23, Gorm, Gr8dude, Grafen, Griffin5, Guinness2702, Guy Harris, Hadal, Hanche, Helpfulweasal, Heron, Hmwith, Hu12, Hutchyy, Hydrargyrum, Iced Kola, Improv, Isnow, JLD, Jarijokela, Jbroderi, Jcarella, Jdulaney, Jeff Song, Jeffq, Jerome Charles Potts, Jerryobject, Jimmayoy007, John of Reading, Johnsonja, Johnuniq, Jojalozzo, Jonathan Oldenbuck, Josemanimala, Jpatokal, Jtjdt, Julesd, Julien, Julo, Justin Ormont, Karl.brown, Katieh5584, KeithTyler, KelleyCook, Kiran Gopi, KnightRider, Koman90, Kostpolt, Krymson, L33th4x0rguy, Leion, LtNOWIS, Luna Santin, MARKELLOS, Mac, Mackeriv, Martarius, MartynDavies, Mauls, MaxiLego, Meleegy, Metageek, Metagraph, Michael Hardy, Mik01aj, MikeKn, Mintleaf, Mitch Ames, Mmsarfraz, Modster, Mojodaddy, Moocha, Moocowsrock, Moraja, Mr Minchin, Mroach, Mushroom, Nanda, Nathan Hamblen, Neelix, Nelhage, NerdyNSK, Nikevich, Nil Einne, Niteowlneils, Nivix, Nokezie, ObuK, Ocram, Ohnoitsjamie, Olivier, Omegatron, Oosoom, Panindra, Pascal.Tesson, Paul August, Paul Marshall UK, Pc123-123, Peter Karlsen, Petertorr, Phantomsteve, Plaws, Plustgarten, Pnm, Ponydepression, Pushp vashisht, Pythoulon, Qxz, Radagast83, RadioActive, Raindeer, Rairden, Ravedave, Rjwilmsi, Romanskolduns, Ronhjones, Rossumcapek, Rubena, Salimfadhley, Sanders muc, Satori Son, Scepia, Sega381, Shadowjams, Shaimay2, Shan2097, Shirishag75, Shubhrajyoti.ece, Sidasta, SilentForce, Silverxxx, Simon Shek, Simonjwall, Sin Harvest, Situwei, Skeejay, Slakr, Sligocki, Slusk, Smalljim, Smyth, Snuffkin, Spa34a, Squash, Starshadow, Stephan Leeds, SteveSims, SudoGhost, Suicidalhamster, SuperMog2002, SystemBuilder, T0ky0, TBadger, Taka76, Talrias, Tapuwiki, Tentoila, Teque5, The Thing That Should Not Be, TheFallenCrowd, Thumperward, Timeastor, TitaniumDreads, Titoxd, Tngu77, Toh, Tom.k, TonyW, Toolnut, Tothwolf, Towel401, Trasz, Tripbeetle, Triwbe, UmbertoCorponi, UncleDouggie, Unixxx, Unyoyega, Utuado, Uzume, Vegaswikian, Velella, Vicentemf, Virgofenix, Watgap, Wfaulk, Whpq, Wideangle, Wikijimmy, Wikiklrsc, Wikiliki, Wingman358, Xompanthy, Xrtc, Yelyos, Yvh11a, ZIKRIE, ZXIII, Zaian, Zarcillo, Zephyris, Zeugma fr, Zr40, 417 anonymous edits

HTC_Flyer *Source*: http://en.wikipedia.org/w/index.php?title=HTC_Flyer *Contributors*: Antonisbouritsas, AvicAWB, Awesomeness95, Crashmatrix, Dj777cool, FamilyGuy1998, GeekyPuppy, Horizonsperson, IGEL, Jim.henderson, JoeBikerul, Jwojdylo, MER-C, Mr Sheep Measham, Pnm, SF007, Sin-man, SoWhy, Sokka54, Srirambms, Theo10011, Woohookitty, Zackkatz, 45 anonymous edits

Crayon Physics Deluxe *Source*: http://en.wikipedia.org/w/index.php?title=Crayon_Physics_Deluxe *Contributors*: 21655, 711groove, Austin Yim, AxelBoldt, Blurpeace, Brianreading, Calaverx11, Chargh, CommonsDelinker, Comrade Graham, Cybercobra, Debby5.0, Drat, Edcolins, Elipongo, Ember of Light, Frankatca, Hamster X, Himan1238569, Ismashed, KieferSkunk, Kindofdavish, Lozeldafan, Lukipuk, MagFlare, Marcus Brute, Megata Sanshiro, Mika1h, Minjabuuu, Moomoomoo, Neurolysis, One-Man Army Corps, Piano non troppo, Pixelface, R'n'B, Randomran, Reedy, Remcovlaanderen, Rettetast, Simeon, Skybon, Slagheap, StealthFox, TDogg310, Tassedethe, Waffle, Xebozone, Xihr, Yuanchosaan, Yzmo, ZeroOne, 77 anonymous edits

Wacom *Source*: http://en.wikipedia.org/w/index.php?title=Wacom *Contributors*: AVRS, Acdx, Aizuku, Akriasas, Al Lemos, Alexsh, Alpha Quadrant, Alphathon, Anomie, Antilived, Arthena, BMB01758, Badmonkey, Bartłomiej Kwiatkowski, Batpox, BeBoldInEdits, Bewildebeast, Bigexplosions, Bongwarrior, Bovineone, Cambridgeblue1971, Cameronbell76, Candelabre, Chameleon, Chazzy88, Chrisphin, CliffC, CommonsDelinker, Craetwin, Cyrius, Dave6, Dismas, DocWatson42, Donlibes, Download, Drakcap, Drilnoth, Droob, EOBeav, Eetwartti, Endroit, Eptin, Fthiess, Furrykef, Gafaddict, Geek 2.0, George100, GeriatricCondition, Gesiwuj, Gobonobo, Googleaseerch, Gordeonbleu, Harryboyles, Hellmark, Humblefool, Ignacioerrico, J o, Jason Stormchild, JasonAQuest, JeffJonez, Jeroentje, Jorbettis, Jovianeye, Ka Faraq Gatri, Kakurady, Kintetsubuffalo, KirbyMeister, Kjlewis, Knowledgexxx, Kwamikagami, Lester, Lindes, Lordmetroid, LrdChaos, Macs79, Mailer diablo, Makitake, Manop, Martarius, Masato araki, Mikael Häggström, Minrice2099, Motherships, Mr Person, MrOllie, NeuronExMachina, Neuroscript, Nikevich, Nz101, OlavN, Paddles, Paul August, Pe2pe2pe2, Pne, Pnm, Prattflora, Quadell, RazorICE, Rbucci, Red, Rossumcapek, RxS, Saladbar1, Sarc37, SaudiPseudonym, Saxbryn, Shack, Sheephorn, Siałababamak, Soccerbri, Sonicfeet, Soren121, Sotaru, Tarale, Tedmund, TheGeekHead, Theshadow89, Thumperward, Top Jim, UCDS, Uber Kosh, Ultima669, Unidue.hilfskraft, VectorCell, West.andrew.g, Whaa?, WiiWillieWiki, WikiLaurent, WisK, Wonchop, Woohookitty, Zachwoo, Zerueu, 222 anonymous edits

Soft_key *Source*: http://en.wikipedia.org/w/index.php?title=Soft_key *Contributors*: Alfpooh, Bearcat, CarbonX, Ericg, Malcolma, MaxPont, Pnm, Romulus99, Widefox, 6 anonymous edits

Image Sources, Licenses and Contributors

File:Samsung Galaxy Note.jpg *Source*: http://en.wikipedia.org/w/index.php?title=File:Samsung_Galaxy_Note.jpg *License*: unknown *Contributors*: User:Rwxrwxrwx

file:Samsung Galaxy Note Screenshot.png *Source*: http://en.wikipedia.org/w/index.php?title=File:Samsung_Galaxy_Note_Screenshot.png *License*: unknown *Contributors*: Rwxrwxrwx

File:Android logo.png *Source*: http://en.wikipedia.org/w/index.php?title=File:Android_logo.png *License*: unknown *Contributors*: Google

File:Android robot.svg *Source*: http://en.wikipedia.org/w/index.php?title=File:Android_robot.svg *License*: unknown *Contributors*: Google

File:Android 4.0.png *Source*: http://en.wikipedia.org/w/index.php?title=File:Android_4.0.png *License*: unknown *Contributors*: Android Open Source project

Image:Android dog.jpg *Source*: http://en.wikipedia.org/w/index.php?title=File:Android_dog.jpg *License*: unknown *Contributors*: secretlondon123

File:System-architecture.jpg *Source*: http://en.wikipedia.org/w/index.php?title=File:System-architecture.jpg *License*: unknown *Contributors*: Jgaliana, MB-one

File:Android home.png *Source*: http://en.wikipedia.org/w/index.php?title=File:Android_home.png *License*: unknown *Contributors*: Unamed102

File:Galaxy Nexus smartphone.jpg *Source*: http://en.wikipedia.org/w/index.php?title=File:Galaxy_Nexus_smartphone.jpg *License*: unknown *Contributors*: Faramarz, MB-one, SF007, 1 anonymous edits

File:I'm Watch smartwatch with Android.jpg *Source*: http://en.wikipedia.org/w/index.php?title=File:I'm_Watch_smartwatch_with_Android.jpg *License*: unknown *Contributors*: Armbrust, SF007, 1 anonymous edits

File:Google_TV_Screenshot.png *Source*: http://en.wikipedia.org/w/index.php?title=File:Google_TV_Screenshot.png *License*: unknown *Contributors*: Tokyoship

File:Androidmarket3311.png *Source*: http://en.wikipedia.org/w/index.php?title=File:Androidmarket3311.png *License*: unknown *Contributors*: Griffin5, SF007

File:AndroidMarketPermissions.png *Source*: http://en.wikipedia.org/w/index.php?title=File:AndroidMarketPermissions.png *License*: unknown *Contributors*: Amlz, SoWhy

File:Android chart.png *Source*: http://en.wikipedia.org/w/index.php?title=File:Android_chart.png *License*: unknown *Contributors*: Android Open Source project

File:Group of smartphones.jpg *Source*: http://en.wikipedia.org/w/index.php?title=File:Group_of_smartphones.jpg *License*: unknown *Contributors*: gillyberlin

File:IBM SImon in charging station.png *Source*: http://en.wikipedia.org/w/index.php?title=File:IBM_SImon_in_charging_station.png *License*: unknown *Contributors*: User:Bcos47

File:Nokia 9210.jpg *Source*: http://en.wikipedia.org/w/index.php?title=File:Nokia_9210.jpg *License*: unknown *Contributors*: J-P Kärnä

File:Htc Touch Pro2 Georgy.JPG *Source*: http://en.wikipedia.org/w/index.php?title=File:Htc_Touch_Pro2_Georgy.JPG *License*: unknown *Contributors*: User:Georgy90

File:Original iPhone docked.jpg *Source*: http://en.wikipedia.org/w/index.php?title=File:Original_iPhone_docked.jpg *License*: unknown *Contributors*: Andrew from London, UK

File:Global Mobile Applications Store Revenue.svg *Source*: http://en.wikipedia.org/w/index.php?title=File:Global_Mobile_Applications_Store_Revenue.svg *License*: unknown *Contributors*: User:MySchizoBuddy

File:Smartphone share current.png *Source*: http://en.wikipedia.org/w/index.php?title=File:Smartphone_share_current.png *License*: unknown *Contributors*: -- Eraserhead1 <talk> 12:49, 3 March 2010 (UTC) Graph created by myself. Original uploader was Eraserhead1 at en.wikipedia

File:IFA 2010 Internationale Funkausstellung Berlin 03.JPG *Source*: http://en.wikipedia.org/w/index.php?title=File:IFA_2010_Internationale_Funkausstellung_Berlin_03.JPG *License*: unknown *Contributors*: User:Bin im Garten

File:IFA 2010 Internationale Funkausstellung Berlin 18.JPG *Source*: http://en.wikipedia.org/w/index.php?title=File:IFA_2010_Internationale_Funkausstellung_Berlin_18.JPG *License*: unknown *Contributors*: User:Bin im Garten

File:-order.gif *Source*: http://en.wikipedia.org/w/index.php?title=File:-order.gif *License*: unknown *Contributors*: Aotake, Sarang, Swift, Wikic

File:-red.png *Source*: http://en.wikipedia.org/w/index.php?title=File:-red.png *License*: unknown *Contributors*: Sarang, Ymursal, Yug

File:Lenovo-X61-Tablet-Mode.jpg *Source*: http://en.wikipedia.org/w/index.php?title=File:Lenovo-X61-Tablet-Mode.jpg *License*: unknown *Contributors*: User:Evan-Amos

File:N800 frontside2.jpg *Source*: http://en.wikipedia.org/w/index.php?title=File:N800_frontside2.jpg *License*: unknown *Contributors*: Zxc

File:IPad in Case.jpg *Source*: http://en.wikipedia.org/w/index.php?title=File:IPad_in_Case.jpg *License*: unknown *Contributors*: Yutaka Tsutano

File:ASUS_Eeepad_Transformer_with_Dock_Keyboard.JPG *Source*: http://en.wikipedia.org/w/index.php?title=File:ASUS_Eeepad_Transformer_with_Dock_Keyboard.JPG *License*: unknown *Contributors*: User:Fieldafar

File:Xo3-fuse-2.jpg *Source*: http://en.wikipedia.org/w/index.php?title=File:Xo3-fuse-2.jpg *License*: unknown *Contributors*: laptop.org

Image:HTC Touch2 used with a stylus.jpg *Source*: http://en.wikipedia.org/w/index.php?title=File:HTC_Touch2_used_with_a_stylus.jpg *License*: unknown *Contributors*: User:Asim18

Image:Styluses.JPG *Source*: http://en.wikipedia.org/w/index.php?title=File:Styluses.JPG *License*: unknown *Contributors*: User:Cspurrier

File:IFA-messen.jpg *Source*: http://en.wikipedia.org/w/index.php?title=File:IFA-messen.jpg *License*: unknown *Contributors*: Ole Morten Knudsen/Teknofil.no

File:Bundesarchiv Bild 183-H10252, Berlin, Funkausstellung, J. Goebbels, H. Kriegler.jpg *Source*: http://en.wikipedia.org/w/index.php?title=File:Bundesarchiv_Bild_183-H10252,_Berlin,_Funkausstellung,_J._Goebbels,_H._Kriegler.jpg *License*: unknown *Contributors*: Florival fr

File:Bundesarchiv Bild 102-00877, Berlin, Eröffnung der Funkausstellung.jpg *Source*: http://en.wikipedia.org/w/index.php?title=File:Bundesarchiv_Bild_102-00877,_Berlin,_Eröffnung_der_Funkausstellung.jpg *License*: unknown *Contributors*: Jcornelius

File:Bundesarchiv Bild 102-00880, Berlin, Eröffnung der Funkausstellung.jpg *Source*: http://en.wikipedia.org/w/index.php?title=File:Bundesarchiv_Bild_102-00880,_Berlin,_Eröffnung_der_Funkausstellung.jpg *License*: unknown *Contributors*: Cobatfor, Jcornelius, YMS

File:Bundesarchiv Bild 102-08307, Berliner Funkausstellung, Riesenlautsprecher.jpg *Source*: http://en.wikipedia.org/w/index.php?title=File:Bundesarchiv_Bild_102-08307,_Berliner_Funkausstellung,_Riesenlautsprecher.jpg *License*: unknown *Contributors*: ALE!, Haabet, Infrogmation, Mattes, Santosga, Shoulder-synth

File:AT&T logo.svg *Source*: http://en.wikipedia.org/w/index.php?title=File:AT&T_logo.svg *License*: unknown *Contributors*: Alpta, Armbrust, Avicennasis, Eastmain, Fairlyoddparents1234, Fæ, Henry W. Schmitt, Jeff G., KansasCity, Koman90, LAX, Mendaliv, Oxymoron83, Presidentman, Sfan00 IMG, TheNewPhobia, Videmus Omnia, Zigger, 11 anonymous edits

File:Increase2.svg *Source*: http://en.wikipedia.org/w/index.php?title=File:Increase2.svg *License*: unknown *Contributors*: Sarang

File:Decrease2.svg *Source*: http://en.wikipedia.org/w/index.php?title=File:Decrease2.svg *License*: unknown *Contributors*: User:Sarang

File:AT&T logo (horizontal).svg *Source*: http://en.wikipedia.org/w/index.php?title=File:AT&T_logo_(horizontal).svg *License*: unknown *Contributors*: Avicennasis, Benstown, Closeapple, Fairlyoddparents1234, Koman90, Sfan00 IMG, Wgabrie, 1 anonymous edits

File:AT&THQDallas.jpg *Source*: http://en.wikipedia.org/w/index.php?title=File:AT&THQDallas.jpg *License*: unknown *Contributors*: FoUTASportscaster

Image:Southwestern_Bell_logo_(1984).png *Source*: http://en.wikipedia.org/w/index.php?title=File:Southwestern_Bell_logo_(1984).png *License*: unknown *Contributors*: KansasCity

File:SBC Communications logo.svg *Source*: http://en.wikipedia.org/w/index.php?title=File:SBC_Communications_logo.svg *License*: unknown *Contributors*: Avicennasis, Benstown, Koman90, MBisanz, Stickguy, The Navigators

File:Att-logo.svg *Source*: http://en.wikipedia.org/w/index.php?title=File:Att-logo.svg *License*: unknown *Contributors*: Avicennasis, Iliketimmyturner, Koman90, Train2104

File:Lockup.svg *Source*: http://en.wikipedia.org/w/index.php?title=File:Lockup.svg *License*: unknown *Contributors*: User:Alpta, User:Andros 1337, User:BetacommandBot, User:Goldfish007, User:KansasCity, User:PNG crusade bot

Image:BellSouth logo.svg *Source*: http://en.wikipedia.org/w/index.php?title=File:BellSouth_logo.svg *License*: unknown *Contributors*: Avicennasis, Fairlyoddparents1234, Koman90, Retro00064, Svgalbertian

File:Att building LA.jpg *Source*: http://en.wikipedia.org/w/index.php?title=File:Att_building_LA.jpg *License*: unknown *Contributors*: User:KennethHan

File:At&tPhone.JPG *Source*: http://en.wikipedia.org/w/index.php?title=File:At&tPhone.JPG *License*: unknown *Contributors*: Brownings, J o, Shoulder-synth, 2 anonymous edits

File:AT&TOrange.JPG *Source*: http://en.wikipedia.org/w/index.php?title=File:AT&TOrange.JPG *License*: unknown *Contributors*: Brownings, Can't sleep, clown will eat me, 1 anonymous edits

File:Randall Stephenson, CEO of AT&T.jpg *Source*: http://en.wikipedia.org/w/index.php?title=File:Randall_Stephenson,_CEO_of_AT&T.jpg *License*: unknown *Contributors*:

File:SER klein exhibits.djvu *Source*: http://en.wikipedia.org/w/index.php?title=File:SER_klein_exhibits.djvu *License*: unknown *Contributors*: EFF

File:Att-midtown-center-atlanta.jpg *Source*: http://en.wikipedia.org/w/index.php?title=File:Att-midtown-center-atlanta.jpg *License*: unknown *Contributors*: Original uploader was Connor.carey at en.wikipedia

File:Texasdd.JPG *Source*: http://en.wikipedia.org/w/index.php?title=File:Texasdd.JPG *License*: unknown *Contributors*: User:Tehumanetwork

File:Red River Rivalry Logo.png *Source*: http://en.wikipedia.org/w/index.php?title=File:Red_River_Rivalry_Logo.png *License*: unknown *Contributors*: NThomas

File:Galaxy S II Logo.png *Source*: http://en.wikipedia.org/w/index.php?title=File:Galaxy_S_II_Logo.png *License*: unknown *Contributors*: User:Hikarugo

File:Samsung Galaxy S II (3).jpg *Source*: http://en.wikipedia.org/w/index.php?title=File:Samsung_Galaxy_S_II_(3).jpg *License*: unknown *Contributors*: User:TechGizmo

File:Samsung Galaxy S 2 and its removable parts.jpg *Source*: http://en.wikipedia.org/w/index.php?title=File:Samsung_Galaxy_S_2_and_its_removable_parts.jpg *License*: unknown *Contributors*: User:YuMaNuMa

File:Samsung Galaxy S2 (3).jpg *Source*: http://en.wikipedia.org/w/index.php?title=File:Samsung_Galaxy_S2_(3).jpg *License*: unknown *Contributors*: User:TechGizmo

File:Samsung galaxy10.jpg *Source*: http://en.wikipedia.org/w/index.php?title=File:Samsung_galaxy10.jpg *License*: unknown *Contributors*: Mumbaifreaks

File:Samsung Galaxy S2 Epic 4G Touch.jpg *Source*: http://en.wikipedia.org/w/index.php?title=File:Samsung_Galaxy_S2_Epic_4G_Touch.jpg *License*: unknown *Contributors*: User:Alvinpingol

File:Samsung Galaxy Tab.jpg *Source*: http://en.wikipedia.org/w/index.php?title=File:Samsung_Galaxy_Tab.jpg *License*: unknown *Contributors*: Closedmouth, Davepape, Eeekster, Eraserhead1, Explicit, WOSlinker, We hope

Image:Nokia E51 Black.jpg *Source*: http://en.wikipedia.org/w/index.php?title=File:Nokia_E51_Black.jpg *License*: unknown *Contributors*: Original uploader was Feci1024 at en.wikipedia

File:IPhone 4S in hand.jpg *Source*: http://en.wikipedia.org/w/index.php?title=File:IPhone_4S_in_hand.jpg *License*: unknown *Contributors*: User:Mungous

File:Samsung Omnia 7.jpg *Source*: http://en.wikipedia.org/w/index.php?title=File:Samsung_Omnia_7.jpg *License*: unknown *Contributors*: StevieBallz (http://www.pocketpc.ch/members/stevieballz.html)

File:SonyEricssonW980-001.jpg *Source*: http://en.wikipedia.org/w/index.php?title=File:SonyEricssonW980-001.jpg *License*: unknown *Contributors*: User:Ofsch

File:BlackBerry Torch.jpg *Source*: http://en.wikipedia.org/w/index.php?title=File:BlackBerry_Torch.jpg *License*: unknown *Contributors*: edans / Enrique Dans

File:N95 Media-keys-open.jpg *Source*: http://en.wikipedia.org/w/index.php?title=File:N95_Media-keys-open.jpg *License*: unknown *Contributors*: User:Asim18

File:Motorola-milestone-wikipedia.jpg *Source*: http://en.wikipedia.org/w/index.php?title=File:Motorola-milestone-wikipedia.jpg *License*: unknown *Contributors*: User:Shritwod

File:Nokia 6710 Navigator (open) (3284598189).jpg *Source*: http://en.wikipedia.org/w/index.php?title=File:Nokia_6710_Navigator_(open)_(3284598189).jpg *License*: unknown *Contributors*: James Nash

File:Motorola flipout.jpg *Source*: http://en.wikipedia.org/w/index.php?title=File:Motorola_flipout.jpg *License*: unknown *Contributors*: User:Philphos

File:Nokia n93-1.jpg *Source*: http://en.wikipedia.org/w/index.php?title=File:Nokia_n93-1.jpg *License*: unknown *Contributors*: affemitwaffe

file:SIM Card.jpg *Source*: http://en.wikipedia.org/w/index.php?title=File:SIM_Card.jpg *License*: unknown *Contributors*: User:Georgy90

file:SIM Card Holder.jpg *Source*: http://en.wikipedia.org/w/index.php?title=File:SIM_Card_Holder.jpg *License*: unknown *Contributors*: Richard Wheeler (Zephyris)

file:Tf sim both sides.png *Source*: http://en.wikipedia.org/w/index.php?title=File:Tf_sim_both_sides.png *License*: unknown *Contributors*: User:Koman90

file:SIM chip structure and packaging.svg *Source*: http://en.wikipedia.org/w/index.php?title=File:SIM_chip_structure_and_packaging.svg *License*: unknown *Contributors*: User:Justin Ormont

file:GSM Micro SIM Card vs. GSM Mini Sim Card.svg *Source*: http://en.wikipedia.org/w/index.php?title=File:GSM_Micro_SIM_Card_vs._GSM_Mini_Sim_Card.svg *License*: unknown *Contributors*: User:Justin Ormont

file:Telia micro SIM with brackets.jpg *Source*: http://en.wikipedia.org/w/index.php?title=File:Telia_micro_SIM_with_brackets.jpg *License*: unknown *Contributors*: User:Mroach

file:Disassembled SIM Card Film.JPG *Source*: http://en.wikipedia.org/w/index.php?title=File:Disassembled_SIM_Card_Film.JPG *License*: unknown *Contributors*: User:Dabombazzz

File:Embedded SIM from M2M supplier Eseye with an adapter board for evaluation in a Mini-SIM socket.jpg *Source*: http://en.wikipedia.org/w/index.php?title=File:Embedded_SIM_from_M2M_supplier_Eseye_with_an_adapter_board_for_evaluation_in_a_Mini-SIM_socket.jpg *License*: unknown *Contributors*: User:Paul Marshall UK

file:Thuraya sim.jpeg *Source*: http://en.wikipedia.org/w/index.php?title=File:Thuraya_sim.jpeg *License*: unknown *Contributors*: User:Towel401

file:Grameenphone SIM Both Side.JPG *Source*: http://en.wikipedia.org/w/index.php?title=File:Grameenphone_SIM_Both_Side.JPG *License*: unknown *Contributors*: User:Auyon

file:Au ic card.jpg *Source*: http://en.wikipedia.org/w/index.php?title=File:Au_ic_card.jpg *License*: unknown *Contributors*: U s e r :

file:NTT DoCoMo FOMA card chip green.jpg *Source*: http://en.wikipedia.org/w/index.php?title=File:NTT_DoCoMo_FOMA_card_chip_green.jpg *License*: unknown *Contributors*: User:Qurren

File:HTC Flyer Front View.png *Source*: http://en.wikipedia.org/w/index.php?title=File:HTC_Flyer_Front_View.png *License*: unknown *Contributors*: Horizonsperson, Mahahahaneapneap, Sfan00 IMG, Skier Dude

File:HTC Flyer BBuy 5th Av jeh.jpg *Source*: http://en.wikipedia.org/w/index.php?title=File:HTC_Flyer_BBuy_5th_Av_jeh.jpg *License*: unknown *Contributors*: User:Jim.henderson

File:HTC Flyer-Back-2.JPG *Source*: http://en.wikipedia.org/w/index.php?title=File:HTC_Flyer-Back-2.JPG *License*: unknown *Contributors*: User:Srirambms

Image:Wacom Logo WhiteType.svg *Source*: http://en.wikipedia.org/w/index.php?title=File:Wacom_Logo_WhiteType.svg *License*: unknown *Contributors*: Macs79, Mechamind90, 1 anonymous edits

File:Wacom Intuos4 Pen Tablet.jpg *Source*: http://en.wikipedia.org/w/index.php?title=File:Wacom_Intuos4_Pen_Tablet.jpg *License*: unknown *Contributors*: User:Jovianeye

CPSIA information can be obtained at www.ICGtesting.com
Printed in the USA
LVOW041559220312

274341LV00004B/121/P